A DESTINY IN DIAMONDS

by
Albert Joris

BOOLARONG PUBLICATIONS

First published in 1986
by Boolarong Publications,
24 Little Edward Street, Spring Hill, Brisbane. Qld.

National Library of Australia
Cataloguing-in-Publication data

Joris, Albert.
A destiny in diamonds.

ISBN 0 86439 011 4.

1. Joris, Albert. 2. Businessmen — Australia — Biography. 3. Diamond industry and trade — Australia — History. I. Title.

338.7'622382'0924

BOOLARONG PUBLICATIONS,
24 Little Edward Street, Spring Hill, Brisbane. Qld.
Design, reproduction and photo-typesetting by
Press Etching Pty. Ltd., Brisbane.
Printed by Apple Printing Co. Pty. Ltd., Brisbane.
Bound by Podlich Enterprises Pty. Ltd., Brisbane.

PROUDLY PRINTED IN QUEENSLAND

Contents

Foreword

I first met Albert Joris at Broken Hill in 1948 when I sought his expertise in the use of industrial diamonds for mining drills. The war years had brought a dramatic change from gems to the manufacture of a complete range of industrial diamonds.

The change to industrial diamonds had been a great challenge for Jules Joris Pty. Ltd., who were virtually the only ones cutting gems in Australia. The Jules Joris Company had been established in 1930 by Albert's father, Jules Joris, who migrated with his family from Belgium to Sydney in 1930.

Industrial diamonds demanded the shaping of diamonds to the finest tolerances, and also the designing and making of dies, jigs, radius generating equipment, projectors of shadow-graphs, special precision lathes, diamond die drilling and polishing machines, all of which became almost a daily challenge.

After the war the business was split up between Jules Joris Pty. Limited which handled the jewellery, Diamond & Boart Pty. Limited which manufactured the diamond tools, and Concut Pty. Limited which was a service company for sawing, drilling, grooving and breaking concrete.

Diamond & Boart Pty. Limited, apart from manufacturing diamond tools, designed and built sawing machines, hydraulic drilling machines, grooving machines, hand, semi and automatic glass working machinery, most of which were first off, not only in Australia, but sometimes the world.

Then came the Longyear Company from America which was established in Australia for the manufacture and supply of wire line slim hole drilling equipment. The Christensen Company came from Utah to manufacture diamond oil bits and oilrig drilling equipment.

The building of a factory in Adelaide to manufacture diamond drills for oil and slim hole drilling was followed later by the building and setting up of a factory in Manila in the Philippines for the manufacture of drill bits, circular and gang diamond saw blades.

''Destiny in Diamonds'' is an inspiring narrative, written by Albert Joris, diamond merchant and businessman, which traces his ambitions and achievements from early boyhood through fifty-three years in the diamond trade in both gems and industrial diamonds.

It follows on two previous publications, ''Stone of Destiny'' published in 1948, written by Ion L. (Jack) Idriess from notes he received from the late Jules Joris.

The second book was published in 1969 and entitled "The Diamond, Stone of Destiny" with an introduction by Albert Joris.

The author describes being one of the first people to handle and advise on the possible division and manufacture of the 426 carat finest white, pure diamond from the Premier Mine, later to be cut and named the Niarchos.

He relates the story of the fine white 90 carat Briolette of India, then owned by Harry Winston, the oldest diamond on record, owned at one time by Queen Eleanor of Aquitaine in 1146-49 when she was on the Second Crusade with her husband Louis VII of France, and later on given by her to her son Richard the Lionheart who took it on the Third Crusade and used it as ransom when he was imprisoned by Henry IV of Austria.

The author writes about his visits to Dr. Williamson's fabulous diamond mine in Central Africa, and his advice on the marketing of his diamonds.

He writes about Hassamali Javeri, the personal jeweller to the old Aga Khan, and the fantastic emeralds of the late Nizam of Hyderabad.

He writes about his selling of industrial diamonds in London, the man-made diamonds of General Electric, A.S.E.A. and De Beers and the technical papers he presented at De Beers Diamond Seminars at Oxford, Tokyo and Manila.

He writes, and gives credit for the fact, that the late Sir William Walkley of Ampol fame, not only was responsible for the money and the effort to drill the first successful oil well in Australia, but also the catalyst which started the real and later successful search for large diamond deposits in Western Australia.

The author describes how the West Australian diamonds could have been controlled and marketed, and a fabulous diamond industry established in Australia.

He states that lack of entrepreneurship caused the lion's share of the action to go to the South Africans, despite the fact that the knowledge and experience was available in Australia.

The predictions made by the author's late father, of large diamond deposits waiting to be discovered in Australia, have come true.

Sir James Foots

CHAPTER ONE

Boyhood in Antwerp

As a young boy it seemed to me that Antwerp had always been home to the Joris family. As far back as the twelfth century our ancestors had farmed the lands nearby and for three generations before me the Joris name was well-known in Antwerp's diamond trade. Antwerp was, and still is, the world's largest centre for diamond cutting. By the time I was in my early teens my father was employing about 40 people in the cutting and facetting of diamonds. His business was prospering and life in general had been good to us.

My parents had a very comfortabe apartment in Antwerp and, on the outskirts of the city, about a half hour's ride by train, we owned a wonderful two-storey house. It was set in four hectares of garden with every imaginable kind of fruit tree. There were gooseberries, apples, pears, plums and cherries as well as chestnuts and hazelnuts. We also grew strawberries and various other berry fruits and produced all our own vegetables.

We were fortunate at that time to have an Army barracks nearby and, with the Commandant's blessing, it was relatively easy for my father to find enough strong willing lads to work in our gardens. Many of the young soldiers were country boys with farming in their blood. The chance to escape the rigours of Army life for a time, the extra pay, the weekends off, not to mention my mother's wonderful cooking, all served to make working for my father an attractive proposition. Needless to say, our gardens were always a show piece and each season produced an abundant harvest.We also kept a few animals. There were usually a couple of pigs, one or two goats, a horse and a few sheep to keep the grass down. I remember I used to breed angora rabbits as a hobby. In one part of the garden my father had built an outhouse with a loft to store the winter feed. Below was a workshop, which every Joris invariably has, with his forge, anvil, vices and workbenches. Ours also had a brick oven for baking bread. The most wonderful smells would emanate from that workshop at weekends when my father and a friend of his would come to do the baking. He was a baker of some repute. I used to watch him slide the dough in when the bricks were white hot and I would hang around to see the finished loaves pulled out. They were large fat loaves, about 30 to 35 centimetres in diameter,

Jules Joris as a young man in Antwerp, the world's leading diamond centre.

sometimes heavy with currants and other fruits. To me it was the most delicious bread in the world.

At one time my father decided to try his hand at wine-making using our own gooseberries. He bought an old press from the local brewer and installed it at home. The recipe, as I recall, came from Mrs. Beeton's Cookery Book. The juice had to be placed in oak barrels and kept cool. At the appropriate time a hole was drilled in the top of each barrel and a small length of tubing was fitted, its free end leading to a bucket of water. In this way the gases released during fermentation could safely percolate through the water without the risk of air getting in to spoil the brew. One had to be careful, however, to install the device before the pressure inside the barrel became too great. Only rarely did my father misjudge his timing.

I recall the only occasion when what must have been a particularly potent batch of wine was left too long with the result that the whole barrel-full erupted with great velocity as soon as the top was punctured. Driven, perhaps, by the heat and humidity of the day, our old ram managed to dispose of a fair quantity of the stuff before we realised what he was up to. With a stomach full of half-fermented gooseberry wine he became unnaturally aggressive and was ready to charge everyone within sight. He might have been quite dangerous had not the wine dulled both his sense of direction and his equilibrium. As it was, he could manage no more than a rather ungainly lurching motion with his legs frequently unequal to the task of keeping him upright. The poor old thing had to suffer the further indignity of having one of my mother's hats jammed over his horns. I have no doubt that he suffered terribly afterwards.

As for the wine, it was invariably excellent. In fact, my father was even approached by a number of people to form a company to make wine on a commercial basis. He was never tempted to do so, however, and he continued to bottle it and put it down just for our own use and pleasure.

It was my parent's custom to go into Antwerp each Saturday to have dinner and go to the theatre, leaving my sister and me in the custody of our maid. Of course, we encouraged her on such occasions to go out and see her boy friend, leaving us with the run of the house. Sometimes our cousins would come over and we would help ourselves to a bottle from the cellar. As children left to their own devices are wont to do, we devised a series of manoeuvres to determine which of us would be saddled with the washing-up. One of our favourite escapades involved a race over a somewhat tortuous and no doubt dangerous course. It required us to scale an outside wall of the house via the ground floor window shutters, to leap over the balcony on the first floor, charge through my parents' bedroom, and slide down the banisters to the ground floor. Last to reach the kitchen was "it".

At the foot of the stairs my father kept a small bronze statue which proved to be our undoing. It was the figure of a blacksmith, bare chested but wearing a hat and holding a large pair of tongs in his hands. It stood about 65 centimetres high until the afternoon my sister misjudged her rate of descent down the banisters and knocked it headfirst to the floor. To our horror the impact caused the man's head to sink completely into his chest leaving only the rim of his hat resting on his broad shoulders. I suppose it gave him a somewhat whimsical appearance but, foreseeing the gravity of our predicament, we were beyond appreciating the comic aspect of his demise. Of course, my father discovered the deed at once when he came home that night. It was not one of our better days.

I managed to get into quite a few scrapes of one kind or another but I suppose they were really no more serious than the usual run of boyhood misdemeanours. In those days I rather fancied myself as a bit of a crack shot with an air gun. At the front of the house we had a sunken rose garden. At its centre, somewhat grandly stood the statue of a naked male. I think it must have been made of cement or something similar. To one side, a little rose garden seat surrounded by a great clump of rhododendrons made the perfect vantage point. My aim was pretty good, if I do say so myself, and one might guess what part of his anatomy became a favourite target. Before long, permanently testifying to the accuracy of my aim, his "what not" became peppered with numerous small indentations so that he took on the appearance of having succumbed to some vile disease. Inevitably, my father discovered the disfigurement and I remember getting the rounds of the kitchen yet again.

During the winter months the house was closed up except for the period over Christmas and New Year. It was a very special time for us. On Christmas Eve, although we were still only youngsters, my sister and I were permitted to accompany the grown-ups to church for midnight Mass. Most families in the neighbourhood attended Christmas Mass, travelling the short distance to church on foot.

Picture taken about 1905 in Antwerp, Belgium, centre of the world diamond cutting industry. Back row centre is Albert Joris, killed in World War 1 and Uncle of Albert Joris the author of this book. Middle row centre Alphons Joris father of Jules who is sitting on his fathers right and in front of Jules is Joseph Joris.

At that time of year the countryside usually had a thick covering of snow so that our footsteps were silent on the ground. After church, at one or so in the morning, you could make out the small family groups here and there travelling noiselessly across the fields to their respective homes with only the bobbing lights of their lamps and torches to mark their progress through the darkness. Once home, the adults warmed themselves with a cognac or whisky and we all had a feast of Christmas cake, pudding, spekulaas and peperkoek. We had an enormous fireplace, it must have been well over 6 feet high. On those cold Christmas nights we gathered around it and exchanged gifts, making our way to bed finally in the early hours of the morning.

New Year's celebrations were also a big event in Belgium, particularly in the Flanders. My parents always held a big New Year's Eve party with many friends turning up from near and far. Often it continued all through the night with some of our guests staying on for breakfast next morning.

I remember those years so fondly. It was an idyllic childhood in many ways, a fact that I was able to appreciate even then. And so it came as some surprise to me when my parents announced their decision to emigrate. They had decided to settle in South Africa, a logical choice, I suppose, given our family's involvement in the diamond business. At the time the South African Government was encouraging the establishement of a diamond cutting industry. Many of my father's friends like the Van Zwams and the

Furstenburgs had already gone out there with their families and had set up business in Johannesburg. And we intended to do the same. That is, until a chance meeting in Antwerp one Saturday afternoon.

My parents had been to see a film and afterwards went to a favourite café specialising in waffles with maple syrup and ice-cream. An Australian couple, with their daughter and a governess, were seated at a nearby table and were having some difficulty making themselves understood. Most people in Antwerp, particularly those in business, speak English reasonably well, but on this occasion the waitress evidently did not. My father offered his assistance and he and my mother ended up joining them and making their acquaintance.

The Australians were from Sydney and had rented an apartment in Antwerp, making it their home base from which to embark on their grand tour of England and the Continent. In those days, you did not fly from one destination to the next, you travelled by ship. And travellers were invariably burdened down with hefty cabin trunks and a large assortment of bags and cases. While touring, an apartment was the logical place to store the bulk of one's luggage. And Antwerp, being fairly central, was a convenient stop-over point for travellers.

My parents became quite friendly with them and they used to come to our apartment and our country place on a regular basis. They persuaded my parents that Australia would be a far better place to settle than South Africa. "For every shilling you earn in South Africa, you will earn a pound in Australia," they said. It was not true, of course. All the people we knew who went to South Africa and set up diamond factories did exceptionally well. No doubt we would have gained financially too had we gone there. But I don't know if our lives could have been any happier.

Once the decision to go to Australia had been made, the preparations were completed fairly quickly. The house in the country was sold, our possessions packed and the Antwerp apartment vacated. We sailed from Antwerp on a German cargo ship with eight other passengers, including the Australian family, travelling first to Cape Town and eventually putting in at Fremantle several weeks later. As we came alongside Fremantle wharf I can remember my fascination at the sight of so many wharf labourers clad in their blue singlets and wide-brimmed felt hats. I had never seen so many felt hats in all my life.

Our Australian friends had promised to introduce my parents to various business people in Sydney and we were assured of assistance finding accommodation and suitable business premises. All my father's diamond cutting equipment had been carefully crated in Antwerp and now lay in the ship's hold. At Fremantle our friends bought the local papers and got their first real news of home for over a year. I think it must have been their first inkling that the recession had indeed reached Australia. A lot of people were going through the hoop. The ship called in at Adelaide and Melbourne and still the news was not good. At each port of call our friends became more and more subdued. Finally, when we arrived in Sydney, they walked off the ship without a word and it was to be many years before we set eyes on them again. So there we were, on the Sydney docks, without family or friends to greet us, and in the midst of a very grave depression. It was the 4th May, 1930.

CHAPTER TWO

The New Country

I have to admire my parents for their determination to make a new life here. Australia was, after all, quite foreign to them and terribly remote from family and friends. Nowadays, one can hop on a plane and travel to Europe so easily. One tends to forget that in the 1930s Australia really was incredibly isolated. For most Europeans the name 'Australia'' probably conjured up vague images of kangaroos and boomerangs, but relatively few in those days would even have known where it was. One of the first things my father did was to go and pay his respects to the Belgian Consul General. The Consul clearly thought we were mad and counselled my father to book our return passages to Belgium on the next available boat. My father was indignant. He had come, he said, to establish a diamond cutting industry and one way or another he meant to succeed.

For the first few days we put up at a rather depressing little private hotel in Elizabeth Street while my mother scoured the ads each morning in the Sydney Morning Herald. We set off one day in response to an advertisement for a flat at Bondi, catching the tram outside David Jones and rattling our way past St. Vincent's Hospital and row upon row of Paddington terraces. It was to be another forty years before Paddington would emerge from its shabbiness to become such a fashionable and desirable suburb. That morning in 1930, however, it was very much a down-at-the-heel part of the city. I passed by there often over the next few years, always intrigued by the area and its inhabitants. To the casual observer like me, the ladies of Paddington, their heads bristling with curlers, seemed always to be lost in intimate conversation across neighbouring fences.

And so Bondi became our first home. In the meantime my father set about finding a suitable work place. He had a letter of introduction to Sid Raper, the famous oarsman who was then captain of Mosman Rowing Club and also a prominent Sydney real estate agent. In Belgium, our family name was quite well known in rowing circles, my uncle Joseph being a leading rower and later a coach with the Antwerp Rowing Club.

Sid Raper was able to find suitable office accommodation at the bottom end of George Street which provided ample space for all my father's equipment with the added benefit of having a strongroom right on the premises. My father was soon installed with his cutting, sawing and polishing machinery and before long had made contact with the various jewellers around town.

Several weeks passed with little appreciable result. Orders were very slow in coming in and my father began to feel concerned. As it turned out, there was a very good reason why the business was not taking off. It seems the premises themselves had acquired a certain gruesome notoriety. The previous tenant had been a furrier who used the strongroom to store his furs overnight and at weekends. The unfortunate man had been brutally murdered one night right there in the shop and his body concealed in the strongroom. The case had caused a sensation at the time and was given much attention in the press. I suppose, quite understandably, people were uneasy about venturing into the building. There was nothing else for it but to move shop. My father found new premises in the Bank of New South Wales building in Pitt Street. These being more central, and thankfully free of any sinister associations, business immediately began to pick up.

Much of the credit must go to my sister, Elsa, who was then eighteen years old and a very attractive young lady. She helped my father tremendously by getting out a very nice letter outlining my father's skill in repolishing burnt diamonds, repairing chipped stones, recutting old-fashioned stones and so on. She called personally on the managers of nearly every wholesale and retail jewellery store around the city, her good looks ensuring she was always well received. The initiative paid off and it was not long before orders started coming in.

Despite the Depression, my father gradually built up a fine reputation in his trade. For many folk in Sydney, however, these years were very hard indeed. It was our habit, as a family, to go down to the beach at Bondi each day for an early morning swim, even during the winter months when at 6 or 6.30 in the morning it was still quite dark. Sadly, many people were in such pitiful circumstances in those years that they were forced to camp down at night on the sand with only rocks for shelter. Many times my mother would rescue some poor soul after a night in the cold and take them home with us for a hot shower and a meal. That's how tough things were. People would even come up to you in the street and beg for sixpence to buy food. I knew of one place off William Street where it was possible to get a three-course meal for one and threepence.

My father paid me the magnificent sum of ten shillings a week in those days and believe me, we did not enjoy the luxury of a simple forty-hour week. After paying for my tram fares and other necessities, I was lucky to have enough left over at the end of the week for a fourpenny milkshake at Bert's Milk Bar. One miserable, cold winter's evening after night school I was approached in Darlinghurst Road by a young woman in a shabby fur coat. She invited me, for a mere ten bob, to accompany her to her flat. By the way of encouragement she held open the front of her garment, revealing a completely naked body beneath it. I remember feeling acute embarrassment at the time, but looking back I am not entirely sure whether it was simply due to my youthful shyness, or to the realisation that, as usual, my pockets were empty.

About that time, Jack Idriess published a book called "Prospecting for Gold". He had been out of work himself for a long period and was only able to acquire pen and paper for his manuscript through the good graces of the management at Angus & Robertson. It seemed a good many people held the view that since there were no jobs to be had they might be able to make some money prospecting for gold. Unfortunately, very few of them knew the first thing about it and quickly lost even the few pounds they started out with. That's where Jack was able to fill a need, for he knew prospecting inside out. His book took off and was very successful due no doubt to the fact that, since times were so hard, people were keenly receptive to almost any scheme promising financial reward, no matter how slender the odds.

There were plenty of shysters around, too. My father had first-hand experience with one particular couple who managed to dupe quite a number of jewellery stores around the city. This is how they worked it. A man and woman, both quite well dressed and respectable looking, came to my father's office one day asking to see some loose diamonds. So my father set out a small parcel of stones for them containing seven diamonds in all. Now seven is a number you don't have to count. You can tell just by looking at them how many there are. The woman made comments about the stones, and, as she did so, put her middle finger over one of the diamonds. It immediately vanished from sight. When she withdrew her hand, only six stones remained on the table. My father noticed at once and became pretty angry, offering, in no uncertain terms, to knock her "bloody head off" if the seventh diamond did not reappear immediately. The fellow jumped up and shouted, "How dare you insult my wife". From my father's demeanour, however, it must have been pretty obvious that he did not mean to let the incident pass without a fuss, and was perfectly willing to defend his property by whatever force was necessary. Sure enough, when the woman opened her fist, the diamond lay concealed in her palm, evidently attached to a piece of gum or something similar. It was cleverly done and perhaps if there had been a large number of stones on the table they might have pulled it off successfully. As it was, they did get away with several rings and various other pieces from a number of city stores before they were eventually caught.

Elsa and my father used to travel to and from the city each day on the Bondi tram. Coming home one night, as they approached St. Vincent's Hospital near Green Park, the driver pulled up to watch a fight going on nearby between a young policeman and three hoodlums. Everyone on the tram just sat there and stared. No one wanted to get involved. My father, however, was offended by the sight. Three against one was just not a fair go. Calmly he handed his briefcase to Elsa, took of his coat, and went over to offer his assistance.

At that time my father was over forty years old but he had been quite an athlete in his youth and was skilled in both boxing and wrestling. He plunged into the thick of it and laid out two of the culprits pretty quickly. He and the constable between them were able to subdue the third and get him into handcuffs. Then, quite calmly, my father got back on board, put on his coat and resumed his seat as if nothing had happened. Meanwhile, in a unanimous expression of approval, the whole tram erupted into a noisy cheering and applause.

That was the sort of man he was. I am sure it was his strong sense of fair play that later saved me from being expelled from school. Very soon after we arrived I was enrolled at Christian Brothers at Waverley to study for the Intermediate. I did pretty well in languages and maths but did not make much headway in English literature where, as I recall, we were doing the Merchant of Venice.

One thing I had never before experienced was the use of the cane to punish misdemeanours and other trivial offences. One of the Brothers in particular seemed to take a sadistic delight in wielding the cane. On one occasion he called out the whole back half of the class for the "cuts" over some trivial matter. He always raised the cane high over his head and brought it down viciously, following through with the stroke like a champion golfer. I thought the whole thing a disgusting and humiliating practice and was determined not to submit.

When it came my turn I quickly pulled my hand away as the cane came down. The teacher was incensed. On the second attempt I grabbed hold of the cane and threw it down. There was a deathly silence. Nothing of this kind had ever happened before. I believe the teacher came very close to striking me with his bare hand and I was more or less braced for the blow. He hesitated momentarily, however, and seemed to think better of it. Then he stormed out of the classroom to see the Principal. The immediate up-shot was that I was sent home.

Fortunately my father reacted sympathetically when he heard the story and defended my action before the Principal that evening. He had always believed, he said, in one's right to retaliate against physical attack. I suppose the incident had the effect of enhancing my position among the other boys but it did nothing to improve my relations with the Brothers. From that point on I think I was branded as an impudent trouble-maker and life became difficult. At the end of term, therefore, my father agreed to let me leave school and come into the family business. I guess my career options at that moment were in any event somewhat restricted. Where else, in the middle of the bleak Depression, could I find work?

It was not to be the end of my formal studies, however, and for the next three or four years I undertook a course in aircraft construction, studying first with the International Correspondence School and later at Darlinghurst Technical College. Ostensibly, the aim was to try to improve my grasp of English, but I must confess the choice was influenced by a long held fascination with all things mechanical and with flying machines in particular. It proved to be a very useful field of study and I learnt much of the practical application which stood me in good stead later in life.

During my school days in Belgium I had been a keen soccer player. We played whenever the opportunity arose, even with a tennis ball if necessary, when space was restricted. And during the summer months in the country I often participated in the junior bicycle races organised in the village. I didn't very often win anything but I was always a keen competitor and gave it my best.

My Uncle Joseph, as I mentioned, was quite a prominent oarsman with the Antwerp Rowing Club. Later, after he retired from championship rowing, he became a successful coach. His is quite a remarkable story. During the First World War he had suffered severe spinal injuries when a German shell exploded in the trenches. There seemed little prospect that he would ever be able to regain sufficient mobility to cope even with normal physical activity, let alone the sport of rowing, which places such a tremendous strain on the back and stomach muscles. He always attributed his complete recovery to his perseverance with an intense regime of Swedish exercise. I have no doubt it was nothing short of miraculous when you consider his war injuries had made him a virtual cripple and left him confined to a wheelchair.

It was through him that I became a junior member of the Antwerp Rowing Club and, even though I was too young to do any rowing, I was sometimes allowed to sit in the back of the boat and do a bit of coxing.

When we came to Australia, I was keen to get involved in rowing again. Sometime during my first year at work, I presented myself at Mosman Rowing Club one Saturday afternoon clutching a letter of introduction from my uncle to Sid Raper. I believe they must have been acquainted through international competition. It said simply, "Dear Sid, this is my nephew. Try and make an oarsman out of him". So Sid was the first to teach me how to hold an oar and how to sit in a boat. I was later instructed by Mick Murphy, one of the senior club members, who took me out in a pair oar and coached me along until I was proficient enough to get into a novice four.

We had an excellent coach in those days by the name of Dave McGoffin. Dave also coached a light-weight maiden four, but having quite a distinct weight advantage over them, we always considered ourselves superior, referring to them somewhat derisively as "jockeys".

I remember one time, in mid winter, I picked two of them up, one under each arm, and dumped them under a cold shower. When Dave McGoffin realised we were becoming a bit big-headed, he was astute enough to teach us a lesson in the best possible way, arranging a race between the two crews. We thought the whole thing a bit beneath our dignity, expecting, of course, to leave them for dead. "It will be good practice for you," he said.

So he lined us up and started us and, sure enough, we pulled away from them until we were a length and a half in front. We could not maintain the advantage, however, and were more than a little humiliated to see them gradually closing the gap. Eventually they were level with us and we were giving it everything we had, pulling our insides out. Still they kept level, and then they inched ahead. In the end they beat us soundly by two lengths. The unthinkable had happened and it was a salutary lesson to all of us. There was obviously much more to winning races than sheer muscle power. It was the way you applied yourself, and how effective your blade was in the water, that governed the boat's speed. Those four young jockeys had earned our deep respect.

Diamond cutting meant sitting down all day long in front of my mill, and after several months of it my body and mind yearned for some strenuous excercise. It was then that I decided to take up amateur wrestling and I went along to Jimmy Deakin's gymnasium in George Street above Adams Hotel. Jimmy Deakin was the professional welterweight wrestling champion of Australia. When I first went to him we were roughly the same weight, around 11 stone and 5 pounds, say between 70 and 73 kilos. I was a pretty strong lad and certainly the rowing had built up my muscles and stamina. But Jimmy could put my shoulders to the mat anytime he chose. He had perfected all the techniques and was able to put his body into such awkward positions that it was like trying to lever a 100 kilo weight. No matter how strong I may have been, pretty soon, with the effort of trying to lift him, my arms got tired and my legs felt like lead. Jim could sense this immediately, it was part of his great skill as a wrestler. He would then quickly put my shoulders to the mat and there was nothing I could do about it. In time however, I guess I did get better at it. At least our bouts would go on longer and longer and I remember, before I left there, Jim and I could wrestle solidly for an hour without a decision.

I met a number of well-known wrestlers through Jimmy, and got the opportunity to wrestle with them on a fairly regular basis. Many of them were champions in their own right. Jack Knight was an Australian amateur heavyweight champion, Jack Simmons lightweight champion, and Jack Murray, or Gelignite Jack, as he was called, was the welterweight amateur champion. Gelignite Jack had the strongest legs of any man I have ever wrestled with and he was probably one of the fastest.

A lot of the international footballers came to Jimmy's gymnasium before the start of the football season to get into condition. It was always my job to wrestle with them although invariably they would be 3 to 4 stone heavier than me. But they were easy meat simply because, although they had great strength, they didn't know much about wrestling and didn't know how to apply their weight advantage.

Later, when Jimmy closed down I went along to Johnny Paradise in George Street. Johnny was the middleweight professional champion of Australia at the time. As a result of his many tussles in the ring, he sported two incredible cauliflower ears and I often wondered how he was able to hear anything through the tiny openings that remained.

Johnny disappeared not long after that, and when next I caught up with him he was in some sort of self-imposed exile at Broken Hill. It seems he had agreed to put on an exhibition match in Adelaide with a friend of his. As part of their performance, and in true showbiz style, Johnny was supposed to be thrown from the ring, go berserk, and then clobber his opponent over the head with a chair. Of course, it was all carefully staged with the chair being specially constructed from light-weight balsa and designed to fall apart instantly at the slightest blow. Somehow the chairs got switched and everything went dreadfully wrong. Johnny's friend was knocked unconscious and he had to have surgery to stitch up his head. Johnny, under threat of law suit, had to hastily reconsider his future.

As luck would have it, he ran into Maurice Mauby and George Fisher who later became a knight of the realm. At that time they were with the Zinc Corporation at Broken Hill and quite keen on the idea of setting up a gymnasium for the company executives. Johnny seemed like the ideal candidate to take charge of the project and he was offered the chance to go up there and look things over. With his expertise he soon had a fully equipped gymnasium in operation and he was employed by the company to put the executives through their paces and supervise their fitness programmes.

He was doing very well for himself and he enjoyed a fulltime salary plus a lead bonus. But after a while he felt he didn't have enough to keep him fully occupied and, with the company's consent, he decided to branch out into the restaurant business. Not that you could really call his eating place a restaurant, for the menu was pretty well limited to the usual "mixed grill", but it did provide the only place in Broken Hill where you could get a meal at any time of the day or night. Johnny ran the place with his wife, a hard-working German girl from the Barossa Valley, and together they made quite a go of it. His clientele was made up of tough mining men who, after several hours at the pub at the end of their shifts, were apt to wander in very much the worse for wear. Johnny didn't take any nonsense from them. At the first sign of trouble he would come out from behing the counter and start taking off his coat. ''By the Christ, I'll come around there and throw you out,'' he'd yell. Such was his reputation that, by the time he had emerged, all the trouble-makers had made a hasty retreat. The sight of Johnny's cauliflower ears was enough for most of them.

I went to Broken Hill many times in the years ahead because by that stage we were in the business of making diamond drill bits. I would always stop by and have a meal at Johnny's to renew our friendship and catch up on all his news. Later on, he bought himself a pub which he ran successfully for many years. He was a real character and I admired him very much for his energy and drive and sheer determination to get ahead. For a man who could barely read or write, his success was all the more praiseworthy.

Right from those early days, and just as I in turn would do, Elsa undertook her apprenticeship in the diamond cutting business under my father's expert guidance. Within a remarkably short space of time activities in the Pitt Street workshop had increased enormously. About that time someone gave my father a piece of very good advice. What he needed, it was suggested, was someone able to look after all the invoices, receipts and other paper work associated with running a business. In 1931 he engaged Mrs. Esther Naughton and what an absolute gem she turned out to be. With meticulous attention to detail and with each transaction beautifully recorded in elaborate copperplate, Mrs. Naughton brought order and discipline to my father's business records. In the beginning she came in for only a few hours a week, and then gradually more and more as work increased until eventually she was there full-time. She stayed with my father for many years and always had the business very much at heart. Dear Mrs. Naughton, much loved and respected by our family passed away in November 1984 at the age of 83.

They were happy days in Pitt Street before the war. We always seemed to have people in to lunch. In the beginning when I was the office junior, Elsa being three years older, people would come in around lunch-time and my father would say, ''Why don't you have a bite to eat with us? Albert, go down to McIlwraiths''. McIlwraith's was a grocery store in Pitt Street which catered to the lunch-time crowd. More often than not I'd go down there and buy sandwiches only to return moments later for more because somebody else had arrived.

Learning diamond cutting from my father was a tremendous experience for me for he had been schooled in the trade by his father before him and knew the business inside out. In Belgium, where diamond cutting is a very specialised art, he had owned his own factory employing up to about 40 people. One of the most important things he taught me was the art of recognising the source of different diamonds. It is crucial for anyone involved in buying uncut diamonds to know precisely what they are dealing with. Certain diamonds, for instance, like those from South West Africa, go down at least one shade in colour when they are cut. They might look nice fine whites to start with but, after cutting, will lose at least 10% in the value. Some diamonds from Tanzania, however, improve in colour after cutting, often increasing up to 10 to 15% in value. Then again other diamonds, especially those from the Congo or Zaire, have a lot of internal stresses which are released and produce a flaw as soon you try to put a facet on. This could decrease the value of the stone by 15% or more, depending on the size of the flaw.

Once a diamond cutter has graded his stones, he has then to make a choice between sawing or splitting. He might even take the benefit of a splitter or a sawer to seek expert advice. On the other hand, if there is a particularly valuable or large stone involved, an expert tradesman might be called to 'open' the stone. 'Opening' a stone simply means putting windows on it, usually by cutting two or more parallel facets. These allow the cutter to see clearly inside and determine if there are any impurities or flaws. In this way he will know whether to split or saw or both.

In Antwerp, all the different steps in the cutting of a stone are highly specialised and performed by skilled tradesmen. First, there is the man who does crosswork. It is his job to put the 'table' or main facet on, then four facets on the top and four underneath. Next there is the brutter who turns the diamond to a round shape on a lathe, using another diamond to cut it. Then there is the man who cuts in eight. He puts four more facets on top and four more underneath, each cut is perfectly even and parallel and at the correct angle in order to get the maximum possible reflection of light. Finally, there is the man who specialises in 'brilliandeering', that is, putting all the triangular-shaped facets near the table and girdle and between all the main facets. By the time it is finished, a brilliant cut diamond will have 57 facets in all, however, some cutters make a tiny extra facet right on the point or culet, giving it 58 facets. Occasionally, and as I said, in the Antwerp diamond trade they all seem to stick to their own specialities. In the same way the man who works on large stones invariably does not cut small ones. One man may specialise in polishing emerald or marquise cuts or baguettes, but he will generally not touch brilliants, and so on.

The most important tool in cutting a diamond, is of course, the disc. Discs are usually about twelve inches in diameter and made of a special, porous cast iron. The manufacture and care of discs used in diamond cutting is a very specialised business. Fortunately, my father had enough sense to realise before leaving Antwerp that although he had a very wide experience in diamond cutting and polishing he did not know all that much about sawing techniques nor the care of discs. He knew that once he was in Australia he could not just ring up a firm and ask them to come and collect his discs and expect to get them back ready and prepared for cutting and polishing. And he knew that he couldn't just run to my cousin and ask him to saw diamonds for him. All these things he would have to do for himself. So he took lessons in sawing and in the care of discs, in grinding and doing the points, and in balancing. In short, he learnt all the necessary techniques and these, in turn, he taught me.

During the First World War my father was taken out of the Belgian Army and sent on special duty to Birmingham in England where he looked after the Kynock and B.S.A. factories. There he supervised the manufacture of diamond tools which, because of the demands of the war, were becoming increasingly important to industry. At the end of the war, on his return to Belgium, he started back in the gem trade, but continued to work to some extent with industrial diamonds, having an association with Fabrique Nationale, the big factory in Liege where they make Browning revolvers, service rifles and shot-guns. In those days they also made cars and motor bikes. His experience with industrial diamonds and the manufacture of industrial diamond tools was to become very important to us later on, especially during the grim years of World War Two. I mention all this because the opportunity my father gave me to learn the trade from A to Z was something that I believe very few others would ever have had. With my father's enormous experience, I would not have wished for a better teacher.

CHAPTER THREE

Copeton Adventures

It was around 1934 or 1935, I suppose, when we started buying rough diamonds direct from South Africa from a licensed diamond dealer called McAdam & Bower. McAdam & Bower were in Johannesburg and they would go to places like Bloemhof, Christiana and Lichtenburg at weekends and buy diamonds direct from the small diggers who, either by themselves or with a partner and several natives worked their small claims. McAdam & Bower knew exactly what we wanted. They bought not only for us but for people in New York, Antwerp, Amsterdam and other diamond centres.

Prior to 1725, India and possibly Borneo were the only known sources of diamonds. In 1725, diamonds were found in Brazil, and in 1866 in South Africa. Then in 1907 diamonds were found in the Congo, now Zaire, in south-west Africa in 1908, Angola in 1916 and Ghana in 1920. Other finds in Sierra Leone, Tanzania, Botswana, etc., came later. In Australia, diamonds were first found in New South Wales in 1851 by Hargreaves, so it is claimed, at Reedy Creek. Then in 1859 the Reverend Clarke discovered diamonds at Burrendong. In South Australia, Dr. Bleasdale found diamonds at Echunga, then they were found in Victoria in the Gippsland area and Beechworth in 1867. More were found at Cudgegong River near Mudgee, at Bingara in 1872 - 1873, at Tingha in 1883 - 1884 and finally at Copes Creek or Boggy Flats in 1875, all in New South Wales. So you see, the discovery of diamonds in Australia goes way back and, perhaps surprisingly, even pre-dates their discovery in South Africa.

Unfortunately there were not sufficient diamonds being found in Australia at that time. But my father had known of the Australian Copeton diamonds long before we ever came to Australia. As a matter of fact, during the Boer War when Antwerp could not get any diamonds from South Africa, diamonds from Copeton, near Inverell, were one of the few sources of supply.

The Copeton diamonds are completely different from any others, in as much as they are terribly hard. They have what we in the trade call 'naats' criss-crossing all over the stone. Basically, naats are seams, just like welding seams and invariably they have a grain structure which runs in a herringbone pattern. Now, when facetting a diamond, you must cut at right-angles to the grain, so that when you strike one of these naats, it becomes virtually impossible to cut the stone. Because of this characteristic, the

Washing plant of the Australian Diamond Fields (Limited) Copeton, Tingha Division New South Wales, 1900

New Washing Plant at the "Star of the South" Diamond Mine (Boggy Camp, Tingha Division New South Wales) Copeton

Copeton diamonds soon became known in Antwerp as 'Can-Ni-Faire', meaning 'cannot be cut'. In fact, legislation was passed requiring all Australian diamonds (and they were the only Australian diamonds, at the time, being found in any quantity) which came on the market to be invoiced and sold as 'Can-Ni-Faire' diamonds. It seems far too many people had lost their shirts on Copeton diamonds; even the experts couldn't cut them. In those days, practically all diamonds were cut into gems for ornamental purposes and industrial diamond usage was basically unknown.

Copeton diamonds were first discovered by the early tin miners who, following Copes Creek down from Tingha in search of tin, came across a number of 'shiny pebbles', as they termed them. They probably suspected they were diamonds but unfortunately they laboured under the common misconception that a real diamond is virtually indestructible. The test they used in those days was to put one of their shiny pebbles on an anvil and hit it with a sledgehammer. As far as they were concerned, if it broke into pieces it was not a diamond. The trouble is that, although a diamond is tremendously hard, and Copeton diamonds are some of the hardest I have ever seen, they are also comparatively brittle. Invariably, with such punishment, those shiny pebbles either shattered into small pieces or shot away like a bullet. Believing they were worthless, the miners used to give them to their children as playthings to keep in a match-box. Eventually someone decided that they should send some samples across to London. When the news broke that those shiny pebbles were indeed diamonds, there was a big rush to Copeton, which was then known by the rather unenticing name of 'Boggy Camp'. There grew up quite a township, and in its hey-day there must have been 2,000 to 3,000 inhabitants along with an assortment of hotels and shops etc.

The Copeton diamondiferous deposits, as in many other parts of New South Wales, occur in tertiary river drifts and cement representing old river accumulations of more than one geological age and lying at various distances from present river channels. Whether the actual source of the diamonds at Copeton, or of other diamonds found in New South Wales, will ever be known, remains to be seen. I know of three different types of diamonds in New South Wales, all of which come from three different sources. The Kimberlitic Pipes where they originate are probably covered over with one or more layers of basalt. One of the most comprehensive books on the subject was written by Tony MacNevin of the Department of Mineral Resources, and is called *The Geological Survey of New South Wales Mineral Resources No. 42 — Diamonds in New South Wales*. Tony has since left the Department of Mineral Resources, but his study of the subject was most comprehensive and his excellently illustrated book contains a tremendous amount of information.

Since the turn of the century, the area around Copeton has been worked by numerous people and some substantial finds of diamonds have been discovered, predominantly in pockets of the tertiary river system. It was not unusual to discover intermittently huge quantities of fine diamonds but these discoveries would invariably always come to a halt.

After the war years, in 1948, we formed a small syndicate called 'Southern Cross Diamond & Tin Mining Company Pty. Limited' in which Eric Miller, Q.C., Bill Cody, Reg O'Brien, my father and I were partners. Tom Heath,

Diamonds from Copeton near Inverell N.S.W.

an old diamond miner from way back, was manager and worked the Old Farm mine. We put a shaft down and an inclined tunnel until we struck a granite bar. The water was getting quite a problem at the time. We did get a number of diamonds and tin, but it was uneconomical, and we decided to pull out. In 1971, the leases were eventually sold to Audimco Limited.

Copeton diamonds are somewhat similar to the diamonds from Bingara, although Bingara diamonds are usually smaller and inclined to be whiter in colour. The other diamonds in New South Wales are generally found in the Cudgegong River area. These diamonds are very much like African stones from around Kimberley in South Africa, although, of course, Kimberley diamonds also vary depending upon the pipe they come from.

It is interesting to note that when Wellington Alluvial, a subsidiary of BHP, were dredging for gold at Wellington they started at the northern end of the old Macquarie River, which is about one mile from the present river. Seventy-five per cent of the diamonds they recovered in their dredges were Copeton diamonds, the other twenty-five per cent were Cudgegong River diamonds. As they worked down stream the proportion began to vary, and towards the end, before they ceased operations, the figures had reversed. In other words, they were then mining seventy-five per cent Cudgegong River diamonds and twenty-five per cent Copeton diamonds.

At Capertee, which is between Lithgow and Mudgee, a type of diamond is found which has very fine striations. This suggests they have not travelled and were formed locally. In quality, they remind me very much of diamonds from Borneo. What is interesting I think, is that all three types can be found in the same area in fairly equal proportions. Doubtless this might have had something to do with the early river system in New South Wales which ran north and south, rather than the present east and west.

The Copeton diamond is quite different from any other diamond I have ever come across, and I have been handling diamonds for over 50 years. Only about five per cent of them can be cut into gems. The others, because they are so very hard, are ideal for industrial purposes. As I mentioned earlier, the Copeton diamonds have a criss-cross pattern of naats or seams, and can be split along those seams quite easily. In fact, it was Copeton diamonds that my father selected when he first taught me how to split diamonds. Copeton diamonds are also remarkably free from impurities. I have heard some people who have made a study of diamonds say that all diamonds have the same hardness. This is simply not true. To polish a facet on a Copton diamond in a place where there is no naat takes longer than on any other diamond I have ever come across. But those that are of gem quality are, when cut and polished, extremely brilliant.

During the war years I did a lot of experimental work on Copeton diamonds for various industrial applications. I had to turn or brute diamonds into cone shapes for use in the Coventry Matrix thread grinding machines. This I did on a precision lathe fitted with a shadow graph, which I made and patented for the purpose. It gave me a magnification of 60 to 100. On this machine I was able to cut the diamond to the required form using diamond splinters. With other diamonds, like those from Africa, the diamond dust produced from grinding a cone was comparatively coarse. Sometimes a small piece would pull out, leaving a small cavity in a very tiny radius which meant the cone had to be recut. I never had this problem, however, with any of the Copeton diamonds I cut, and their finish was always as bright as if the stone had been polished.

Another remarkable feature of Copeton diamonds is that, when heated above 750 degrees, they do not graphitize like other diamonds. I carried out an experiment once in a graphite tube furnace which I had made, and through which I passed clean and dry hydrogen. I selected a two carat Congo cube diamond and a .45 Copeton diamond and heated both in the furnace. This I did by lighting the hydrogen and bringing the temperature up to 1,350°C. I kept it at this temperature for two hours then I let the furnace cool down to room temperature, at all times with the hydrogen on. The temperature was recorded with an accurate optical pyrometer. The Congo cube diamond, although retaining roughly its original shape and physical size, now weighed only .9 of a carat. There was no loss of weight or shape in the Copeton diamond however.

The fact that Copeton diamonds do not graphitize is obviously the reason why they are able to stand up better than any other diamonds when used for abrasive purposes. During the war we were able to supply Lithgow small arms factory with Copeton diamonds for use on their grinding wheels. There it has been estimated that temperatures of up to 1000°c. occur at the point of contact between the diamond tip and abrasive wheel.

After the war, Copeton diamonds also proved most effective in the blast hole drill bits used at Mt. Isa, and we had a standing order to supply around 500 a month. As you can imagine, the centre diamonds received the brunt of the work because they were virtually forced through the rock, there being almost no rotational speed at that point. When the bits came for salvage, the Copeton diamonds were still standing up proud, whereas all the others were flat down to the matrix.

Copeton diamonds, because they have relatively no internal stresses, lend themselves very well for use as wire drawing dies. But a Copeton diamond will take at least twice as long to pierce using a laser beam. The only disadvantage with them when used to dress abrasive wheels, is that generally they do not have sharp points.

When we first came to Australia, my father, already knowing the reputation of and having had some experience with the Copeton diamond, was dead keen to go and have a look and see where they came from. We arrived in Sydney on 4th March 1930 and within a month or two my father bought a second-hand car, a Studebaker, which we repainted a rich blue and called 'The Bluebird'. That Christmas the whole family packed into the Bluebird and we headed off for Glen Innes, Inverell and Copeton. Some neighbours of ours who were English migrants and who had never been further afield than Parramatta, told Dad, "Once you get into the country, there is no way you'll get petrol". With that warning in mind we set off with two four gallon drums of petrol strapped to the running-board of the car. As the December temperature soared, so our cans started to swell, and my father quickly produced a knife and stabbed a hole in each to avoid the risk of explosion. The petrol spurted everywhere and we were probably very lucky not to have it burst into flames. We poured as much as we could into the tank and proceeded on our journey minus the gerry-cans.

From our point of view it was a fascinating trip for we had not seen anything of Australia apart from Sydney and its surroundings. I did just about all the driving, although I was only fifteen and did not at that stage have a driver's licence.

At Glen Innes we went out to Kings Plains and saw Miss Sach's mine. She produced some beautiful pieces of quartz crystal, which later I had to saw into slabs 3mm thick and 1 inch square for AWA. It was no mean feat, I can assure you, since quartz crystal is very hard.

It was on that trip that we learnt to cook damper, and boil our billy by the camp fire. We generally stayed at good country hotels along the way. In those days, bed and breakfast, and by breakfast I mean a hearty meal of bacon and eggs, toast, jam and marmalade, came to the grand sum of two and sixpence a night.

Apart from one annoying habit, the old Studebaker went very well. Every twenty miles or so we would have to stop, gingerly unscrew the radiator cap, and with a final flick dive for cover to avoid the spray of scalding water that gushed forth from its innards. Then we had to wait around and let it cool before adding additional water. It wasn't until later that I found that a pin in the drive shaft which operated the water pump had sheared off. Once I replaced it, everything went well.

The day I turned seventeen and became legally entitled to be in charge of a motor vehicle on a public road, I presented myself at the police officers in Macquarie Street where the testing of applicants for a driver's licence was conducted. A policeman took me for a drive through the Domain. The first thing he said was, "You've done a bit of driving before, son, haven't you?"; there was no denying it. I had, in fact, clocked up many thousands of miles by then.

It was soon after I got my licence that we heard that a diamond had been found up at Stanthorpe near the Queensland border. It was reputed to weigh over eight carats. Dad told me to get out the car and get there as soon as possible to see if I could buy it. I drove pretty well non-stop and as fast as I could. Sure enough it was a beautiful octahedron diamond weighing 8.25 carats. The fellow I bought it from, after we concluded the deal, offered to let me bunk down for the night and rest before returning to Sydney. A couple of hours later two fellows arrived in a car and said, "We hear you have a diamond for sale and we've come to make an offer on it". "You're too late fellows, this young man has bought it already," he said. So, triumphantly I went back to Sydney and showed it to Dad, who was delighted with it. It was slightly yellow in colour and had a small impurity in it, just above the girdle or centre of the stone which meant we had to saw it just above the four points of the octahedron. We finished up with a large stone and a smaller one and did very well out of the deal.

Later on I used to go to Copeton quite often to buy diamonds from various people. There were only a few huts there apart from the small general store which seemed to be the focal point for the small community. It was run by Tom Govers and his wife. Tom was one of the old-timers at Copeton, and the store had become a sort of meeting place. At night, very often, a number of the locals like Tom Heath, Clem Hill, Tom Cant, and so forth used to gather down there. They would all sit around and tell stories about the diamonds that were found at the turn of the century, and I was eager to learn as much as I possibly could. We would sit by the fire in the winter months in May, June and July when it was as cold as charity. I slept in a little hut a few hundred yards distant from the store. They would always finish up by telling some blood curdling stories about diamond merchants who, having come to Copeton to buy diamonds, would be found next morning with their throats cut, their diamonds and money gone. Of course, it was all made up for my benefit, but I didn't take any risks. There was no lock on the door, but I always made sure that I had a chair securely wedged against it from the inside before I went to bed. The cold was unbelievable, and I also made sure that I had a ready supply of Sydney Morning Heralds to put on the stretcher where I slept.

From Copeton I sometimes used to walk up to the old ghost town of Howell. In the past, the old Howell, or King Conrad Mine, had produced lead, silver, zinc and copper. In the 1890s despite the fact that miners' wages were no more than about one pound ten shillings per week, the mine wasn't paying and was forced to close. Howell had been a fair-sized town, with two hotels, two general stores, a couple of butcher's shops, two bakers, café, barber shop, blacksmith, and even its own newspaper called the 'Digger'.

The old mine was reworked again in about 1899 by Peter Alwell, and intermittently at various times after that. In the days when I used to visit there, however, there was only old George Monkton left.

In his youth George had ridden with Thunderbolt the bushranger. Thunderbolt was eventually shot at Uralla, and George Monkton was taken captive. Because of his youth, however, he was spared the hangman's noose.

I remember one day making the long up-hill trek to Howell to visit old George and have a yarn with him. The ceiling of the hut where he lived was lined with rough hessian and George was sitting at the table having his meal. Suddenly, there was movement in the ceiling above; obviously there was a rat there. George, with a swift upward jerk of his arm, and without a moments hesitation jabbed his fork through the hessian in the vicinity of the movement. A squeal indicated he had hit his mark and speared the rat on the other side of the hessian. Then, quite calmly, he sat down and continued eating using the same fork. He didn't even bother to wipe it. Well, as you can imagine my stomach turned over and I was just about sick. The odd characters I met around there really were a rough and ready mob.

Another well-known Copeton identity was a fellow by the name of Friday. Friday was not his real name, nobody knew his real name, but he arrived in Copeton on a Friday and so that's what they called him. I remember he spent his time between two little huts, one in which he used to sleep, and the other which he used as his kitchen and living area. On the shelf he kept a small medicine bottle, which was always a quarter to three-quarters full of diamonds. Occasionally I used to go and see him to try and buy some of the diamonds, and often I would catch him having his breakfast. He would sit at the table with a loaf of bread and jars of jam and honey, perhaps a lump of butter and cheese, and would cut himself thick slices of bread, liberally spreading it with whatever food he happened to have on hand. Nearby he kept a .22 rifle handy, just in case a rat took an undue interest in his food.

Old Friday was a damned good miner, but he was quite an extraordinary sight and I often wished I'd had a camera in those days to capture him on film. Often I'd find him digging and working in a pair of ladies' milanese bloomers. When I questioned him as to why he should choose such a strange outfit he explained that it only cost two bob and, what's more, they were very cool and kept dust out.

Friday was not a very big chap, but he had the reputation of being a bit of a Casanova with the girls. He came down to Sydney one time with four hundred odd carats of diamonds, and called in to see my father. Dad made him an offer but Friday wasn't satisfied with the price and did not accept. We found out later he had been corresponding with a lady who had placed a newspaper advertisement to the effect that she wanted to meet someone of her own age with a view to matrimony. It appears Friday had made arrangements to meet her on a certain date at Taylor Square. At the appointed time, Friday arrived for the rendezvous and there was a woman waiting on the corner. He said to her, 'Are you Mrs. so-and-so?'' ''Yes deary,'' she said and without further ado took him to a flat which she had close by. Of course, she was not his intended at all but just someone who happened to be there at the right time. The next thing was that the woman's husband came home. He gave Friday the rounds of the kitchen, took his money, took his diamonds, and turfed him out.

A couple of days later two fellows appeared in our office and said they had some diamonds to sell. When they opened the parcel, Dad immediately recognised Friday's diamonds. He looked them over pretending not to have seen them before. In very guarded Flemish, he instructed me to go and fetch the police. When the detectives arrived they arrested the men, and Friday got his diamonds back. But we heard not a word from him, not even a 'thank-you' for what we had done. We heard he had gone off to Melbourne and sold that particular parcel of diamonds to Rene Van der Kelen, who was then the Honorary Belgian Consul in Melbourne. Van der Kelen was a diamond merchant and also in the pearl business. In fact, at that time he was known as the Pearl King. After that, our Copeton Casanova met a barmaid in Melbourne who, when she noticed his roll of pound notes from the sale of his diamonds, became very pally with him. Out they went and did the town, spending his money as if it was going out of fashion. The moment his last quid had been spent, however, she turfed him out too. He hitched-hiked all the way back to Copeton I believe, quite a distance, particularly in those days when there were very few cars or trucks on the road.

In 1935 a family by the name of De Stoop migrated from New Caledonia and settled in Sydney. De Stoop is a Flemish name and the family could trace their origins to an area in Northern France which, in years gone by, had been part of Flanders. My parents met the De Stoop family not long after they arrived and pretty soon Maurice De Stoop, who was then just out of school, started work as an apprentice in my father's business. His first job was to sit behind my mill and try and look through a magnifying glass. In time I taught him how to facet diamonds. Not long afterward his brother Ronnie started with us also, and he sat behind my sister's mill, learning the trade in the same way. Maurice was quite a character, and we used to have a lot of fun in the workshop, but both boys were serious students and later on became top-class tradesmen. They just seemed to have the natural ability and special aptitude to pick it up. It is a skill which I think one must be born with. Some people have it and others don't.

About the time the De Stoop brothers started with us, my sister Elsa met her future husband August Dale. August was a French woolbuyer who was sent out here by the very reputable French firm of Henri Caulliez. In those days, woolbuyers were a rather elite group in Australia, they earned big salaries, travelled all over the continent, and were highly respected.

Elsa chose a suitable diamond, which my father had recommended, and cut a most magnificent emerald cut diamond for her own engagement ring. I think she must have been one of the first female diamond cutters in the world at that time.

They were married in 1937, and later honeymooned in Europe. By then I had saved up some money, and was due for some holidays, so my father let me go to South Africa. I went by cargo ship, which took just 12 passengers and I had a really marvellous trip to Durban. During the voyage, I had befriended a mother and daughter by the name of Turnbull who, after a visit to Australia, were on their way back to England via South Africa. In Durban I bought a second-hand car for one hundred and fifty pounds which the three of us shared on a 4000-mile trip all around South Africa. It was a fantastic experience for all of us. We drove to Johannesburg, then south to some of the diamond diggings at Kimberley, Bloemhof, Christiana and Lichtenburg and on to Vereeniging on the Vaal River. From there we visited the famous Kruger National Park in Northern Transvaal, staying overnight in kraals along the way. At each kraal we had to check in before 6 p.m. or else pay a fine and could not leave before 6 a.m. the following morning. They provided accommodation and a central restaurant where we were able to get all our meals. During the day we were free to drive around to look at the wild-life, and there were plenty of lions, giraffes, elephants and hippopotami wandering free in their natural habitat.

Visitors were not supposed to get out of their cars, although we did break the rules once or twice to have picnics alongside our car. There was an element of risk involved because a lion, leopard or some other animal could do a lot of harm. The authorities naturally discouraged the practice, not only because of the danger to human life but also because of the potential harm to the reputation of the park. One of our big mistakes was to pick up a tortoise we found on the road and put it in the back of the car. Believe you me the smell was absolutely incredible. Within a very short space of time, we had to open the door and abandon the poor creature to its fate.

Whilst in South Africa, I had the opportunity of visiting numerous diamond cutting factories, and with our agents McAdam & Bower, I went around to various diggings, which made it very interesting for me. I was lucky, really, because the Joris name opened many doors for me. When people in the trade heard the name they'd say, "Ah, son of Jules". My father's reputation was known far and wide from the Antwerp days.

At the core of the diamond business is a relatively small group of people, somewhat like an exclusive club where everybody knows everybody else. If you have done the right thing all your life, people in the trade know about it and will respect you for it. Put a foot wrong or do something underhand, however, and the news will spread like wildfire. If the misdemeanour is bad enough, you can become an outcast almost overnight, and you might as well get out of the business.

It was during that visit that I first met Sir Ernest Oppenheimer of the De Beers Company along with one of his top men by the name of Ted Brown. Ted was somewhat of a human dynamo. He had started off as a sorter in Kimberley, or rather an apprentice sorter, which meant that if anybody dropped a diamond off the bench, he had to get down on his hands and knees and look for it. He gradually worked his way up through the De Beers hierarchy and became Sir Ernest Oppenheimer's right-hand man. Ted was responsible for various research projects and also developed Boart & Hard metals, a manufacturer of the tungsten carbide drills used in mining as well as diamond drill bits and numerous other things. Ted was unbelievably energetic. He was a big, hard-working, hard-playing South African. I was convinced, even then, that if Ted ever retired from the De Beers Company they would need six people to replace him. Ultimately, that is just about what happened.

I felt myself very much at home in South Africa, because most of the South Africans I met were very much like Australians, easy going, generous, certainly not pretentious, and generally very easy to get on with. The fact that I understood and spoke Afrikaans also helped a lot.

At the end of my stay in South Africa, I took the car back to Durban, sold it, lost fifteen pounds on it, and boarded a cargo ship for Europe. There I landed at Antwerp and again bought a second-hand car which took me all over Europe and England.

It was wonderful tour for me. In spite of my Belgian ancestry, fate had somehow determined that I should be born in England. It was to England that my parents were sent during the First World War when my father's diamond expertise was needed to assist in the manufacture of various tools. I was born in 1915 at the home of Lady Peel at Ashford Bowdler, near Ludlow in Shropshire. Ludlow is an historical town with a 900-year-old castle, and was one of the border towns where, in the Middle Ages, the English tried to subdue the Welsh people. It was a nostalgic journey for me and whilst there I visited the historic little chapel where I had been christened on the land of Lady Peel.

Back in Europe in the summer of 1938, I met up with my sister and her new husband in Belgium and drove them all the way to Marseilles where I put them on the boat bound for Australia at the end of their honeymoon. They had arranged for me to meet a friend of theirs in Paris. As it turned out he was a native of Paris and knew the city backwards. At his suggestion, we teamed up with another friend, a Swiss ski-ing champion, and the three of us booked into a cheap little hotel in the Montmarte district. From there we set out every day at 10 or 11 in the morning, visiting all the notable museums and places of interest, lunching at little cafés and lazing away the afternoons. In the evenings we sampled a different restaurant each night, before doing the rounds of the famous Paris nightclubs, finishing up in bed around 3 or 4 in the morning. The whole day, including the cost of taxis, lunch, dinner, drinks and nightclubs, averaged out at about 25 francs per day, which was very cheap indeed.

From Paris I went back to England and sailed from Liverpool on the Blue Funnel Line returning to Australia via the Cape. Once home, it was back to the old routine in our workshop, cutting and polishing diamonds.

There were some notable people in the jewellery trade in those days whom I remember very well. My father claimed they knew more about diamonds than most other people in Australia. First, there was Percy Marks, the old Percy Marks, who knew diamonds backwards. He used to give us quite a lot of work in recutting old-fashioned cut diamonds, chipped diamonds, burnt diamonds, and so on.

Another identity was Monty Goldstein who was a regular visitor to our offices. Monty was an inveterate gambler. He used to go to the races every Saturday, and could not resist the Gee Gees. Knowing that we worked on a Saturday afternoon, and knowing that my father used to keep a few pounds in the safe, he would often come running in and say, "Jules, I've got a wonderful diamond here, a wonderful piece of jewellery, a real bargain. I'd like to sell it to you very cheap." On those occasions it was pretty obvious Monty needed money very, very badly, no doubt because he had lost his shirt at Randwick. Most times my father was able to get quite a bargain and did well enough out of it.

Another very well-known identity was Arthur Toombs. Toombs was in charge of the repair and manufacturing workshop at Hardy Brothers in Castlereagh Street. He was highly respected and knew his business backwards. Today his son Geoff is probably one of the most highly respected personalities in the jewellery trade. He is a former President of the Gemmological Association of Australia, and is very knowledgeable, certainly taking after his father.

Then there was Reg Clements. Reg represented Triefus and was a most fastidious English gentleman who had his office in Martin Place. I'm sure he cleaned his diamonds from morning till night. Every diamond he showed us was absolutely spotlessly clean, and what's more he knew colours and purities, and was highly respected in the community.

Another notable person we met during the '30s was Jack Rosenzweig. His name was actually Jacob Rosenzweig and he had left his homeland in Austria at the time Hitler began making things very difficult for Jewish people. His decision to settle in Australia was a wonderful thing for this country because Jack was a superb tradesman and one of nature's gentlemen. He did a lot of work for my father in those days before the war. When war broke out poor old Jack tried to change his name to Jack Roberts. He even went to the trouble of changing all his invoices and letterheads but for some reason fell foul of the bureaucracy and could not get official sanction for it. But, as I said before, he was an asset to Australia and a wonderful tradesman. He trained many people in the art of jewellery making.

CHAPTER FOUR

The War Years

We began work on industrial diamond applications in a small way during the 1930s. It must have been about 1936 or so that I met Tom Massey who set up Century Storage Battery Company in Australia. He was an English-French-Canadian, and quite an inventor. I remember he designed and built a machine, one time, for the printing industry where large sheets of zinc had to be turned perfectly flat and smooth. He hit upon the idea of using a diamond tool and asked me to cut a diamond for the job with its four corners facetted and its underneath turned to a ninety degrees included angle cone. The idea was that you could use one side, and when that eventually wore down, you could undo the clamp in the toolholder, turn the diamond around ninety degrees and clamp it up again and so on all the way round. The diamond that my father selected for the job was, without doubt, the hardest stone I have ever come across. Later on, Tom exported that machine to New York and when the diamond wore out he tried to have it recut. But there was no-one there who could do it. They all found that particular stone too hard. I had only managed to cut it in the first place because I was lucky enough to have a large Indian diamond which I used as a guide. The Indian stone was facetted all over and was about the same size and shape as a sugar coated almond. I prepared a ring on my disc, with this comparatively soft Indian diamond, and in the same ring put the very hard stone. I also fitted a large pulley on the counter-shaft and increased the disc speed from 2,200 to 2,800 revolutions per minute. That was the only way I could possibly cut it.

Around the same time we were approached by Alex Dalgleish from the Tool Equipment Company. This firm makes thread gauges, taps, dies and so forth. In the years before thread grinding came into its own, if you wanted to cut threads in mild steel, brass or copper, you had to use a piece of high speed steel, anneal it by heating it until it was cherry red, and then let it cool slowly, turning it to size on a lathe so that the threads could be cut and the flutes ground. Having done this you then had to heat it again and quench it in oil to harden it. But this process had its drawbacks because it meant that the outside always cooled before the inside making the tap slightly warped, or bent, not excessively so, but enough to cause slight inaccuracies in trying to cut accurate threads.

The modern practice is to turn high speed steel approximately to size, or slightly oversize, then harden it and grind it to size. After it had been ground it is put on a thread grinding machine. This may be either a single rib abrasive wheel like a Jones & Lamson, or a multi-rib wheel like the Coventry Matrix. The ribs on such wheels are produced by a diamond, and it was a tool for this purpose that Alex Dalgleish brought to us to see if it could be made in Australia. Unfortunately, as I did not have the right equipment, I could not do it at the time. It wasn't until later, when I designed and built a shadowgraph which fitted on to a very high precision diamond cutting machine or diamond cutting lathe, that I was able to do it.

People began to bring more and more industrial diamond tools to us. Prior to the outbreak of war, we could actually see from the various industrial inquiries we were getting, that something was brewing. In Europe, the French were still behind the Maginot Line and the Germans were sending across the occasional bomb, but there was a tension gradually building up, which even in Australia we could sense. More and more people were contacting us, saying, ''Can you do this, and can you do that?''

When eventually the war broke out, the Ministry of Munitions came to us and told us we were to move to larger premises and to forget about gem diamonds for a while. We were not allowed to touch them at all. A new bank, the Commercial Bank of Australia, had been built diagonally across from the Pitt Street branch of the Bank of New South Wales, and they moved us into it, to occupy the whole of the second floor.

We were the only ones in Australia at that time who were able to make a full range of precision industrial diamond tools. In the beginning, almost every day, we would get blueprints in from various munition factories, which were then springing up like mushrooms, requesting all sorts of special diamond tools which we had never seen before. Very often we would scratch our heads and wonder how on earth we could do it. Fortunately, my years of mechanical training at technical college, and the fact that I had my own little workshop at home, as well as perhaps my natural inclination to do mechanical work combined with my father's extensive experience, served to give us a good start.

During the war a huge number of machines came in from the United States under lend lease. A lot of the tools essential for their operation needed industrial diamonds for them to perform. Without them they would have been virtually useless. Thread grinding and hardness testing were some of the other applications for industrial diamonds. There were Jones & Lamson thread grinding machines, and even the Sociéte Genévoise machines from Switzerland, which I think had been flown across Portugal somehow, before coming to Australia.

We also made Rockwell hardness testers, another first for us, which required specially cut round diamonds for testing the hardness of hardened steel. Other precision jobs included the making of extremely accurate pyramid-shaped diamonds for Vickers hardness testers and ball-shaped diamonds for testing the hardness of paint. Paint testing was done by spraying the paint on to a piece of flat metal and drawing it beneath a diamond ball set into a tool loaded with so many grammes of weight. The grooves were then measured to determine whether the paint was the right texture, hardness and consistency to be used on warships.

During the war more and more tungsten carbide, which was then a new power metal alloy, was used for turning tough steel, particularly for use in shells. So we had to manufacture diamond impregnated wheels, both resinoid bond and metal bond. This, again, was completely new, but I managed to buy a second-hand hydraulic press in North Sydney which had electrically heated plattens on it and a capacity of about 35 tonnes. The press was hand operated. We had to make dies and obtain bakelite powders, mix them with diamond dust, press the ring and put a backing on top, and then press and cure in order to produce a resinoid wheel. Finally, this would be put in a lathe and turned to the exact bore size required. The same process was applied to metal bond diamond wheels where up to three or more different powder metals had to be mixed with diamond dust and either cold pressed or hot pressed. The face of the wheel containing the diamond dust would then have to be mounted on to a backing of either steel or aluminium and turned on the lathe, before it was ready for use.

We probably achieved a few firsts in those days and it was only well after the war had ended that people started talking about grinding steel by using diamond impregnated grinding wheels. Throughout the war years, however, we had supplied diamond wheels to the Lithgow small arms factory for grinding the hardened steel cocking pieces of rifles.

We were lucky enough to find some excellent tradesmen during that period. One of the first we had was a Frenchman by the name of Emile Colomb. Emile had been knocking around the islands for a number of years working as a ship's engineer. He was one of those naturals, although I didn't think he had any formal training. He had the happy knack of being able to pick things up just by watching people, and it seemed he could do almost anything. Give Emile a piece of fencing wire and I think he would turn out a most elaborate machine for you.

Emile always took a great pride in the jobs he did. In his battle to become Australianised, he had picked up a lot of Australian swear words, which he mixed with French and scallywag swear words he had picked up somewhere in the islands. This private language of his he used quite effectively with a very strong French accent.

When Emile made a steel die for the powder metal section, he would spend hours finishing it and lapping it, and he would really come up with a first-class job. It would hardly have left his proud hands when, on one of those occasional days when everything goes wrong, a young recruit would pick up the beautiful die and wreck it in a couple of minutes. Emile would fly of the handle with a, "You bleeding Kanaka barbarian Australian! I spent hours making that die, you spent minutes wrecking it! You—!" The whole workshop would listen entranced while Emile foamed on into the strangest-sounding, most fiery, of all swear languages.

The boys facetting the diamonds used to sit in front of the windows facing on to Pitt Street in order to get the best light. I well remember the day when one of them got two pennies, heated them over the gas flame, and threw them into Pitt Street below. At that moment, a high-ranking Army officer happened to be going by with his wife. He heard the pennies clatter to the pavement and stooped to pick them up. Well, you can imagine what he said! Murphy, the police sergeant on point duty at the corner of Pitt and Market Streets, was soon on the scene and came straight up to the office, asking for my father. "Mr. Joris," he said, "some of your fellows have been heating

Picture taken during World War 2, when Jules Joris Pty. Limited cut industrial diamonds for the Ministry of Munitions. Picture shows Jules Joris and Albert Joris, August, 1941 Sydney.

up pennies and dropping them out of the window''. Dad, looking very innocent said, ''Sergeant, my boys wouldn't do a thing like that''. Not much they wouldn't!

At the back of our building there was a lane, and on the other side of the lane was Sargents. Sargents were very well known in those days for their meat pies. They had a catering business too, with a restaurant backing on to the lane way. Once a week the garbage people used to come with a truck and back into the lane, picking up large cans of garbage which came mostly from Sargents restaurant. One day, two fellows were busy loading garbage on to the back of the truck when one of our boys, with an extremely well-aimed tomato, hit one of the men behind his ear. He thought, of course, his mate had done it and there ensued a fairly colourful bun fight. All manner of things flew backwards and forwards along the lane way. The walls and windows were splattered with rotten vegetables, and all the while our boys were hanging out of the windows cheering on the hapless combatants.

The bank manager was a tall man with very red hair. He was a very nice bloke but resented the fact that he was obliged to house a factory on the secondfloor of his brand new bank building with lathes, drilling machines, milling machines, shapers, welding equipment and heaven knows what else. He hated the idea of the boys going down to McPherson's and returning with grease-covered lumps of steel and then travelling in his brand new lift.

More than once, I'm sorry to say, he was the target for a water bomb from the second-floor windows.

One of the jobs we did in those days was to set single-pointed diamonds for wheel dressers in powder metal before silver soldering them into steel shanks. When these dressers came back, we used to turn the powder metal back as far as we could before touching the diamond, and then put all the little nibs containing the diamonds into an acid solution in order to burn them out. We didn't have any suitable fume cupboards for this purpose and the chap in charge of this work hit upon the idea of putting the beakers out on the window ledge. This ensured that all the smelly acid fumes were kept out of the workshop.

As it happened, the floor above us was occupied by a German watchmaker. I remember his coming down to see my father in a very agitated state, saying, "Joris, I ran away from Germany because Hitler wanted to put me in the gas chamber. What are you trying to do to me?" The poor fellow had good reason to complain. Apart from the offensive smell, you can imagine what the fumes did to his delicate watch parts.

In the latter years of the war, it became apparent that it was impossible for the Australian government to import diamond wire-drawing dies. A diamond wire-drawing die is a diamond with a hole drilled through the centre of it, with a cone-shaped entrance and exit. Fine wire such as copper, tungsten, nickel or any other wire suitable for such a specialised purpose, is drawn through the hole in the diamond, reducing it in size until extremely fine wire is produced. Most of the wire-drawing industry, which is very specialised, was centred in Belgium, Holland and France, and when war broke out a number of trained men from over there went to England and the United States to set up the industry in those countries. However, both England and America were so urgently busy trying to supply their own requirements that it was out of the question for them to send any wire-drawing dies to Australia. We had to look after ourselves.

One day, a couple of important people from the Ministry of Munitions came to the office and said, "Right, you now must make wire-drawing dies". I enquired, "How?" They replied, "Oh well, we have a couple of Professors, one in Canberra and one at the CSIRO. They can tell you how to go about it."

I saw these chaps, both good men in their own line, and very willing to help, but they knew almost as little as I did about this particularly obscure, highly specialised job. So I worked still later into the night, trying to design a little machine that would do the job. Again I approached the Ministry of Munitions and explained that with more and more jobs of increasing intricacy, such as the making of precision wire-drawing dies, I would have to have the services of some tool or instrument makers, highly skilled men. I was told this was no problem whatsoever, and people would be sent down to help me make these diamond wire-drawing dies.

I don't know just how the misunderstanding occurred, but I still remember how flabbergasted I was the next day when along came a cheeky little bunch of bright-eyed girls, all fresh from school, and all excited at the idea of working in what they called "the Diamond Factory"! They had all admired diamond engagement rings in the jewellers shops, but not one had ever seen a micrometer, not one knew what its uses were, let alone how to use it. So, with all the work on hand, I now had to train these young schoolgirls to read micrometers, to handle precision instruments, eventually to drill holes through that hardest of all stones, the diamond, holes that had to be utterly accurate and finer than a human hair. I have helped train many an excellent diamond-tool technican since, both men and women, but I am still amazed at the speed and uncanny skill with which these young schoolgirls learnt the trade. Previously, after two unsuccessful attempts, I had finally perfected a machine capable of drilling a hole through a diamond. We built a number of these machines and started production.

The first dies we turned out were not marvellous, but they did the job. Gradually the product and quality improved until with experience we were making a reasonable wire-drawing die. My father encouraged me warmly during those experiments, but it used to make his heart bleed when he saw how we were using beautiful, gem-quality diamonds that he knew would make beautiful jewels. The war was on and these gems had to work as tools, hard-working tools, in the production of the fine wires so essential to just about every instrument: in aircraft, telephone sets, in electric-light bulbs, in all the conduit and leads that run through a house, and in motor cars. All that wire had to be produced in Australia and could only be produced by being drawn through wire-drawing dies. Now we knew what was behind the words of the Ministry of Munitions in those grave days: "You just *have* to produce diamond wire-drawing dies!"

My father used to try and buy whatever diamonds he could lay his hands on, whether they came from Copeton, from Beechworth in Victoria, or wherever, but it was a precarious business because in war time diamond production here had virtually ceased. So we were almost entirely dependent on South Africa, and our agents McAdam & Bower.

McAdam, being the canny Scotsman he is, would never insure any parcel of diamonds he sent out to us but would carry the insurance himself. After going through a compulsory inspection in Johannesburg, (South Africa has a 15% export duty on diamonds), the parcels would go by air to Cairo, which was a hot bed of intrigue and theft. From there they travelled to Bombay which wasn't much better, then on to the Cocos Islands, Perth, Adelaide and Melbourne, before reaching Sydney. Yet no parcel was ever lost or stolen, which is an absolute miracle to me. Mac would ring up Lloyds and find out what the premium was for sending a parcel valued at so many thousand pounds to Australia. They would give him a price and he would deposit that amount in a special interest bearing account, keeping his fingers crossed that the diamonds would reach Sydney safely. We would cable him the moment they arrived. By the end of the war he had a tidy little nest-egg bearing interest for him.

When I think about it now I can see that, with the exception of a few men from the Ministry of Munitions and from the munition factories, very few people realised the importance of industrial diamond tools during the war years. The millions of pounds of machine tools that came into this country under lend lease could not have been used without precision diamond tools which we produced.

At the beginning of the war I tried to enlist twice, once in North Sydney and once in Darlinghurst. After three years at technical college learning air craft construction, I had visions of getting into the Air Force with a commission. I rather fancied myself wearing a nice blue uniform with a smart cap. But it was not to be. When I was asked what sort of work I was doing, a quick telephone call to the Ministry of Munitions soon put paid to that.

We worked long hard hours during those years, often clocking up fifty-six hours per week before overtime. It was tedious work, sitting as I had to, in a dark room from morning till night, watching a screen and cutting diamonds to fit into a particular pattern. It played havoc with my eyes, so much so that I had to take a week off on doctor's orders to go to Craigieburn at Bowral. If you were to sit all day in a picture show, week after week, month after month, I think you will begin to appreciate how I felt.

Then every so often I had to go and visit the munition factories at Lithgow, Bathurst and Orange or I would go to various firms in Newcastle, Melbourne, Adelaide and Whyalla, to offer advice, where possible, on the better use of industrial diamond tools. Lithgow was a very big user of industrial diamond tools, and so were the factories at Footscray and Maribyrnong in Victoria. One of the useful things that I was able to show the boy in Lithgow was a method I had devised for bending and twisting tungsten carbide tips without breaking them. It saved a tremendous amount of grinding using diamond impregnated tools.

I used to enjoy going to the small arms factory at Lithgow and I got to know a lot of people there. In fact, quite a few worked for me later on, people such as Vic and Frank Vassella, Jack Stewart, Tim Taylor and others. They were really a race apart, all highly skilled trades people, toolmakers, turners and fitters. There were some real characters there too, and they could be a bit tricky at times. If they made a gauge and it happened to be a couple of tenths of thousand under size, they would put it in dry ice and then bring it back to room temperature. This meant it grew back what it was before, plus a couple of tenths, and Meteorology would then pass it as being spot on.

Jack Stewart was in charge of the big Sociéte Genévoise jig borer, and being a typical example of Swiss engineering, it was absolute perfection. One of the things that fascinated me was how it got to Lithgow from Geneva during the war years. After all, it was a very large machine and it would have cost in those days at least a quarter of a million pounds. Jack looked after that machine with the same care and attention as someone who had just bought a brand new Rolls-Royce. He would clean it and polish it constantly.

Jack was fond of playing tricks on people. On those days a lot of workers used to carry small Globite bags with their lunch inside. They used to put them on the parquet floor alongside their lathe or milling machines, whatever they were working on. Jack could never resist it when he saw one of these Globite cases standing in the passageway. He would take a run and give it an almighty kick, sending it miles along the parquet floor. Anticipating this, someone had taken his lunch out one day and substituted a couple of lumps of steel. Jack took one kick at it and you can imagine what happened. I don't know whether he broke his big toe or not, but it took him a long time before he was able to walk properly again.

And they got their own back on Jack in other ways. As I said earlier, old Jack used to polish that machine like nobody's business. He would never stop for lunch or knock off in the evening until every last steel shaving had been wiped away and the machine was spotlessly clean.

Vic Vassella and Duncan McDiamond were possibly the best turners that ever came out of the Lithgow small arms factory. Both of them could make a lathe talk. Vic had been the Australian motor bike champion at one stage, and I remember once buying a car from him. It was a Desoto Airflow, and I think it was the best car I've ever owned. The finished lapping of the valves was done with cigar ash so that they shone like mirrors. The cylinder head and the cylinder block had been hand scraped so it only needed a piece of tissue paper rather than a gasket to stop any water or oil from escaping. Every nut and bolt was chrome plated. All of this had been done, of course, with the compliments of the Lithgow small arms factory.

The war years were a tremendous education to me because, as I said earlier, we were the only ones in Australia who made a complete range of industrial diamond tools. Because we were breaking new ground, we also had to draw, design and in many cases patent various jigs, attachments and other machinery in order to produce those tools. There was no one to whom we could turn when problems arose because everybody was too busy solving their own design problems.

The only diamond tools we didn't make at that time were diamond drill bits and reamers which are used in the mining industry. There was already a firm in Melbourne making those things and the Ministry of Munitions said, "They make drill bits and reamers, you make precision diamond tools" and that was it. Of course, that didn't stop us making them the moment the war was over.

When eventually the war did come to an end, the Commercial Banking Company wanted to get rid of us. The didn't want a factory in the middle of Pitt Street, and certainly not in their new, or comparatively new, bank building. We didn't want to leave because we didn't know where to go. The Ministry of Munitions had by then lost all its clout and was consequently of no assistance to us. So we had to fight the bank in the courts. They won in the end and we got our marching orders. We were at a loss at that stage to know where to turn next.

It was a friend by the name of Maurice Carrier who came to the rescue. Maurice worked for Kreglingers, a large Belgian wool firm, and he had many useful business contacts in the shipping world. Through him we eventually found premises at Stanmore on Parramatta Road. It was a large storehouse which was then used by one of the shipping companies. They kindly allowed us to move in and use the upper floor, the ground floor being covered with chains, blocks and tackle, timber desks and all manner of things used on ships. Eventually we took up more of the bottom floor until the shipping company finally moved out and we occupied the whole building.

It was a great disappointment to me that the Australian Government was not far-sighted enough when the war was over to begin to stockpile industrial diamonds. Diamonds have always had the reputation of being a girl's best friend, but we looked upon them as the crown jewels of industry. The war years had been living proof of their vital role in industry. I remember going to Melbourne and pleading with Colonel Thorpe, among others, for Australia to prepare itself in this way for any future national emergency. They would not have had to go through diamond dealers like ourselves or anyone else, they could have dealt direct with the South African Government or with De Beers. It would simply have been a matter of buying up diamonds in small quantities over a period of time and putting them away in a vault in the Commonwealth Bank. Diamonds don't go bad, don't rust, don't eat anything and don't occupy much space. But they do appreciate in value. I tried very hard to persuade the powers that be that such a project was a vital necessity. But, as I say, the war was over then, and whilst the officials in the Ministry of Munitions listened sympathetically, they did nothing, and the whole thing came to naught.

The United States were the only people to pursue such a course and they stockpiled all sorts of strategic materials, including industrial diamonds. I hoped all along, the same as my father did, that somewhere, somehow, diamonds would be found in substantial quantities in Australia. Not so very long ago, of course, that became a reality. But that is a story for later on in this book.

CHAPTER FIVE

A Sportsman Travels to South Africa

Even before we left the Pitt Street workshop, we had started experimenting with making diamond drill bits. At that time we were casting them in berryllium copper which is an extremely tough alloy believed to have been used for tools by the ancient Egyptians about three thousand years ago. To make the bits we used a graphite mould, turned to the correct radius inside and out and drilled with tiny holes where the diamonds were to go. We had to apply suction beneath the mould because diamonds have a specific gravity of 3.5, and the moment we poured molton copper in, the diamonds would float to the surface like corks. We had to hold them down somehow, and suction was the only means we could think of to solve the problem. It proved to be quite satisfactory and we ended up with beautiful looking berryllium copper bits set with diamonds.

Our first orders came from John Mc D. Royal who was a drilling contractor in Castlereagh Street. He was also responsible for guiding us as to the various types and sizes of bits required by industry following Canadian and American standards.

Although these copper bits looked beautiful, they did have one significant disadvantage. The fact was that they didn't withstand abrasive action as well as bits make from powder metal such as tungsten carbide. By then we were already manufacturing diamond bits for the mining industry for such people as Broken Hill South, Broken Hill North, the Zinc Corporation, Mount Isa Mines, as well as the Department of Mines in South Australia and Queensland. But we didn't know how to go about making the tough tungsten carbide bits that other manufacturers overseas were producing. When I went to South Africa in 1945 at the beginning of a round-the-world trip, it was with the clear intention of discovering how the process worked.

In Johannesburg, I went to see Ted Brown, Director of the De Beers Company and also Industrial Distributors (1946) Limited, Boart & Hard Metals, and various other subsidiary companies of the Anglo American Corporation. Then I went to the Diamond Research Laboratories which were being run by the De Beers Company for the benefit of diamond tool manufacturers. The Diamond Research Laboratories which were then under the able direction of Dr. Jan Custers and Mr. Cook Weavind. Both were most helpful, giving me a great deal of information on diamond orientation, crystallography, and the effect of fluorescence on the various types of diamonds. They also explained quite a lot about the sub-sieving of micron diamond dust. But when it came to the manufacture of diamond drill bits, thier mouths were sealed. This work was being done by Boart & Hard Metals, and nobody was allowed into their factory. They were unable to tell me anything about the manufacture of powder metal diamond drill bits. The secrets of the various manufacturers' methods were well kept.

One afternoon Ted Brown of De Beers rang me to say he would like to see me the following morning. When I went round he told me that they had found a very large diamond and that Sir Ernest Oppenheimer had asked for my opinion on the stone. Sir Ernest was a very astute man, and whenever a large diamond was found it was his habit to call in two, three or more people to get their opinions. These opinions would help him form a better idea of the value of the stone and allow him to compare his own judgment with that of other people. Knowing I was in town and knowing also that apart from handling industrial diamonds I had been cutting gem diamonds from boyhood, he sought my opinion on the big diamond.

Needless to say, I was on pins and needles when I arrived at Ted Brown's office the next morning.

The diamond in question was produced and it was the most beautiful I have ever seen in my life. It was roughly the shape and size of a matchbox, weighing 426¼ carats, and the finest blue-white I have ever seen. It was an irregular blocky-shaped diamond. After examining it for quite a long time I could not find any imperfections in it whatsoever, except a very small surface flaw at one of the long ends where it did not matter anyway.

We didn't have the money to buy a diamond of that size and quality, but since I had been paid the compliment of being asked for an opinion, I had to try and decide what I would do with it if it was my diamond. The stone could be cut into a huge marquise cut, or into a very large emerald cut. But the market for a stone of that size and colour is so limited that there would probably have been only a few people in the world able to afford it. The alternative would be to divide it and make into two, three or even four smaller diamonds. I called for and was given fine drawing pens and Indian marking ink and up to lunch time I drew lines where I would split it, and marked the part I would saw off. I was then reasonably happy with the tentative divisions I had made.

After lunch I called in Ted's secretary and asked her if I could have a couple of large potatoes and a large knife. She was a little taken aback at my request since I had just had lunch. With the sharp knife, I laboriously shaped several raw potatoes into the shape of the diamond. I then cut the potatoes along the lines where I split and cut the diamond, and where I would saw it. This gave me a good practical idea as to what the diamond would look like after it had been split and sawn. It also gave me a rough idea of the weight before cutting and polishing. My task took the rest of the afternoon and, believe me, I was tired when 5 pm came along and I had to hand the diamond back. After making a few brief notes I took the potatoes and some of the drawings I had made back to my hotel room.

The next morning I sat down and wrote a lengthy report on how I would split and saw the diamond, on what size the rough would be before cutting, and on what I estimated the diamond would weigh after cutting and polishing. I gave my various reasons for splitting and sawing the stone in these particular directions, and supplied drawings to go with the suggestions. Then I gave an estimate of what I thought the finished diamonds would fetch, of the approximate time it would take to split, saw, cut and facet the various stones, and also of the approximate labour costs involved in doing the work. I took my report, drawings and cut potatoes back to Ted Brown, who handed them over to Sir Ernest Oppenheimer. Later on that afternoon I was informed that Sir Ernest wanted to see me at 9 am the next morning.

When I went along he had the report in front of him and obviously had been reading and studing it. He complimented me on the work that I had done. "Excellent! Obviously you have spent a lot of time doing this and I am very, very grateful." He then said, "Perhaps you would like to buy it?" But I explained that we didn't have that kind of money, much as I would have liked to have bought it. So he said, "All right, what can I do for you in return?" Believe me, it didn't take me long to seize the opportunity. I told him I would very much like to have a look through Boart Products and see how they made powder metal diamond drill bits. He thought for a while, then he said, "Right. You can spend the next week or fortnight if you wish at Boart Products. You can take any notes you like, make any drawings you wish, but please, no photographs!"

During the next fortnight I went to Boart Products from 9 in the morning until 5 in the afternoon, and secured all the drawings and details that I so urgently wanted. These I put into a report which I sent to Sydney. The result was that when I got back to Sydney after my travels through Europe, England and the United States, our factory was already producing powder metal bits.

The 426¼ carat diamond was eventually sold by the Central Selling Organisation to Harry Winston. Harry had it cut in New York by his chief cutter, and I am very proud to know that the diamond was divided, split and sawn along almost the identical lines I had marked on the stone originally. This diamond was found in the Premier Mine, where the famous Cullinan diamond weighing 3,126 carats was found many years ago.

The main diamond was cut in a pear shape with 144 facets. It was sold by Harry Winston to the Greek shipping magnate, Stavros Niarchos, and was named after him. But the big diamond also yielded an emerald cut weighing 40 carats and a marquis of 30 carats. Sir Ernest Oppenheimer was a very knowledgeable man when it came to diamonds, and he considered this diamond to be the most perfect colour of any diamond he had ever seen, irrespective of shape or size.

Harry Winston really made a name for himself in the jewellery trade. Harry started off in the jewellery trade by manufacturing cheap engagement rings which sold in those days for about $60 each, which he sold in gross lots to some of the big American retail stores. Later on, he started buying larger stones and some of the better class jewellery. This led him to setting up a whole building in Fifth Avenue, New York. He also owned a shop in Paris and he had offices in Geneva and Rome. After that Harry Winston did exceptionally well dealing with some of the more famous diamonds.

In the New York building which Harry owned, he had a series of rooms, one containing nothing but sapphires and diamonds, another containing nothing but rubies and diamonds, another rubies alone, another emeralds,

The Niarchos diamond.

and one with nothing other than diamond jewellery. Harry was a pretty shrewd operator. He not only sold jewellery, but also hired it out to some of the small jewellers who wanted to set-up shop. They could go along to Harry Winston and hire pieces of jewellery for display in their windows, paying not only rent but also the insurance premium on them. If the jeweller was able to sell any jewellery on hire Harry would pay him 10 per cent commission on the sale. Harry also used to rent out jewellery to debutantes who wanted to look their best at their debut.

Once my wife and I were in Geneva, I rang Harry and asked whether I could visit him with my wife to have a look at some of his famous jewellery and diamonds. His answer was, "Certainly, by all means". After the usual greetings Harry asked his secretary to get out a few diamonds for my wife to look at. Slowly she produced a box, with a least twelve stones in it each over five carats. Harry being a real showman said, "Oh take that away and show Mrs. Joris something worthwhile". So she took the box away and came back with a tiara. The centre stone must have been at least 20 or 25 carats, with large stones set all around. I remember my wife couldn't believe it when she saw it. It belonged, I believe, to one of the Spanish queens. Then she brought out a sapphire, which was the most beautiful colour that I have ever seen, weighing 120 carats and surrounded by large diamonds (about a carat each) in a brooch. Then he said, "How about showing Mrs. Joris that special". So away the secretary went and came back with a 90 carat diamond, in a pear shape, which had been facetted all over. On the pointy end of the pear shape, there were two holes which had been drilled into the diamond at an angle so that a golden loop could pass through and the diamond could be worn as a pendant. The diamond had originally been cut in India, and had been refacetted later on. It was one of the finest clean white stones that I had ever seen. Harry told me he wanted $1,000,000 for it, which was in those days an awful lot of money. When I asked him whether he would be able to sell it, he answered, "No problems, no worries".

When he told me the history of the diamond, I thought it sounded like a pretty far-fetched tale, but then when I researched the matter, I found that Harry's story was true. The diamond concerned is called the Briolette of India. The first known owner was Eleanor of Aquitaine, who probably acquired it in Asia Minor where she took part in the Second Crusade in the years 1143 to 1149, together with her then husband Louis the 7th of France. Eleanor gave the diamond to her son Richard the Lionheart, who took it with him on the Third Crusade. I can't imagine Richard the Lionheart wearing a 90-odd carat diamond underneath his steel armour whilst fighting in the Crusade, but apparently this is what he did. When Richard the Lionheart returned to Europe he was imprisoned by Henry the 4th of Austria, in a castle somewhere on the Danube. Nobody knows whether he used the diamond as a ransom, or whether he was recognized because of it, but he eventually managed to escape.

The diamond then disappeared for a while and reappeared around about the 16th century when Henry the 2nd of France gave it to his mistress Diane de Poitiers. She can be seen wearing it in some of her portraits, which are at Fontainebleau today.

The stone then again disappeared, only to reappear again in 1950 when an Indian maharajah sold it to Harry Winston.

One of the things that fascinated me about the stone was the fact that the technology existed at the time to enable them to drill two holes at an angle such that they would meet at the pointy end of the stone. Having great difficulty in drilling holes through diamonds for diamond wire drawing dies, I realised the skill and time it must have taken to do just that.

Life was indeed hectic in our Pitt Street workshop during the war years, but there was still time for a little romance in my life. At a party one night given by our neighbour's daughter, I was introduced to a beautiful young lady called Margaret Nichols. Years earlier, when I was still a bit of a kid at school, I had walked the route between tram stop and school gate with Margaret's elder brother Bill.

Of course, neither of us had any inkling in those days of what would eventuate in the future. It was not until 1945 in fact, when I was staying in Johannesburg, that I realised Margaret was the girl for me. I cabled her and asked her to be my wife. We were married in Sydney on 9 November that year at St. Mary's Cathedral. It was a decision that neither of us have ever regretted.

In those days petrol was still being rationed and coupons were rather hard to come by. But my old wrestling partner, Gelignite Jack Murray managed to pull some strings and rescue the situation. He got us a tank full of petrol somehow and we were able to take off for our honeymoon at Craigieburn not far out of Sydney at Bowral.

Towards the latter part of the war years I had rejoined Mosman Rowing Club. It was in pretty bad state at that time and its ranks had been depleted somewhat with so many boys away on service overseas. With so few rowers left, the club had gone to rack and ruin. It owed a bit of money around town and the club house was almost derelict.

So I thought, "Well, the only way we can get this club going again is to try and get a decent four. We need David McGoffin to do a bit of coaching for us. We might even win a race or two if we're lucky." Fortunately, a young lad called Bob Tinning turned up there one Saturday afternoon with some impressive experience as an oarsman at the GPS Regattas. We put Bob in the stroke seat, with me in the three seat, Jock McKenzie in two and Bill Purcell in the bow seat, with Dave McGoffin coaching us. We did quite well with nine wins, two seconds and one third out of twelve starts. Of course, in those days we didn't have the money to transport our boats to regattas. We had to row ourselves there and back. It meant that, if there was a regatta on at say Sydney Rowing Club, we had to row all the way up the Parramatta River, then race, and row all the way back again. I can assure you that getting past Circular Quay, with waves being kicked up by ferries, wasn't much fun. It was very hard work.

But the club prospered. We managed to get it back on its feet again by organising a dance every month, and sometimes we were able to get one or two of the Tivoli artists to come and put on a show for us. I remember it used to cost us two guineas a time for their services, with a good supper thrown in as well.

I was captain of the club for about 18 months during that period but I had to leave in the latter part of the war years to spend some time in South Africa. After my marriage on my return to Australia, things at the club were not the same somehow. There was a bit of factional fighting going on and consequently some tension in the air. Bob Tinning, Jock McKenzie and I decided to leave. Bob went off to join Sydney Rowing Club and Jock and I went to North Shore Rowing Club. There, Spencer Grace, the Olympic oarsman, was a prominent figure in the club and was forming an eight to row in the New South Wales Championships. The championships are held over a three mile course on the Nepean River. Our eight had Don Read, a very experienced man in the coxswain seat, with Phil Caizer, Bruce Goswell, Spencer Grace, me, Jock McKenzie, George Wolstonehulme, Ron Moody and Harold Wilks in the bow seat.

We used to row every morning before heading to work at 9 am. Margaret and I had not long been married so, as you can imagine, I was not very popular. I used to sneak out of bed at about 5.30 am, quickly make my bride a cup of tea, and then head off to Lane Cove for the morning session. On top of that we trained two nights a week on Tuesdays and Thursdays, plus Saturday afternoons and Sunday mornings.

The three mile course in fresh water on the Nepean River, particularly when there is no movement in the water, can be a very gruelling race. I am very glad that they have since reduced it to 2000 metres. Of that crew, both Phil Caizer and Bruce Goswell went to the Olympic Games in Helsinki, and so did Bob Tinning, who was then rowing for Sydney. Bob Tinning later became one of Australia's leading orthopaedic specialists. I decided to give rowing away for a while after the championships and do the right thing as a young husband. But then one day I ran into Jack Partridge on the train and was persuaded to join the Mosman Club again and row in the veterans eight. They used to row each Sunday morning for a couple of hours from about 8 o'clock and so I became a regular participant again.

Then about the middle of January, we would always hire a couple of Halvorsen cruisers and set off for a marathon row from Bobbin Head to Windsor and back. It was a distance of about 200 miles all up, not as the crow flies, of course, but measured over the winding course of the Hawkesbury River. We'd split into two groups, one on each cruiser, along with a skipper and a skipper's mate. The skipper's mate also doubled as an emergency rower in case somebody got so many blisters on his hands, backside or feet that he needed to pull out and rest for a day or so. I had devised my own means of avoiding that fate and made a point of dipping my hands in the salt water which, incidentally, runs all the way through to about Wisemans Ferry. I found it a sure means of hardening and toughening my skin so that I didn't end up with too many blisters.

The trip used to take us about 9 days, and each boat was loaded with good food, wonderful steaks and a lot of beer and wine. We would get up pretty early in the morning and, since there was no one else around, just drop our pyjamas and dive over the side for a swim. Then we would put on our rowing togs, launch our boat, and row steadily for about 3 or 4 miles before breakfast. We fell into an easy routine of brisk rowing followed by spells relaxing on board the cruisers, eating, drinking and listening to the occasional news broadcasts. It was a superbly relaxing retreat from city life despite the occasional hard physical effort involved.

The second trip, unfortunately, was marred by a very sad accident. We were on our way back. The day had been oppressively hot since early morning. Ken Preshaw, a friend of mine, was in the stroke seat and I was in the seven seat. We used to row for about 20 minutes and rest for 5 before rowing off again. On that occasion, we had only been rowing for about 10 minutes at the most before Ken called out "Easy oar". So we stopped rowing and he sat stock still. Suddenly he started to keel over to one side. At first I thought he was going to poke his head in the water to cool off, but suddenly I realised there was something dreadfully wrong and I grabbed him. He was dead. Apparently, Ken had always had a bad heart, but he never let on to any of us. He seemed to be superbly fit, playing squash and hockey and sculling nearly every morning at the Spit. It was a sudden and dreadful shock to all of us. We transferred Ken's body to one of the cruisers and were able to stop a speed boat and get a message to Wisemans Ferry to call the police and an ambulance. It marked the end of our rowing expeditions because most of the wives began to worry about their husbands going out after that.

It was only later on that I got involved with rowing again, not as a oarsman this time but as a coach for Riverview. My two sons John and Bill had both been to school there. John had rowed in the first fours at Riverview at one stage, and so when they invited me to take on a bit of coaching, I was happy to oblige. I must confess I thoroughly enjoyed it. After a week at the factory with all the pressures of business and worries about sales, budgets and heaven knows what, it was a sheer delight to be out on the water again coaching the young lads from my vantage point in the speed boat.

In the first year I coached the third and fourth fours. They were a nice bunch of lads, eager and keen to learn. Occasionally, I would jump in the boat and row with them to see how they were going and to give them some practical tips on what to do.

I remember one time one of the young lads complained about a pain in his stomach, so I offered to replace him with another boy until he felt better. "Oh no, I'll be all right Sir, I'll be all right," he assured me. So we took off up the Lane Cove River, all the way up to Lane Cove Park, a bit over four miles in all. Suddenly the lad was overcome with pain again and we had to help him out and put him in the speed boat. At that stage I hadn't rowed for quite a while, not seriously anyway, but guess who took his place in the boat and rowed all the way back again?

The rewarding thing about it all was the fact that we won the odd race, or got a second or third. The next year I was asked to coach the first and second fours, which I did, also with some success. In the third year I coached the second eight. I am very proud to say that in each of those three years, at least one of my crews gained a place at the annual GPS Regatta. Sometimes even both gained a place.

Finally, in the fourth year, I coached the first eight, and I'm glad to say that they won the GPS Regatta. Even more importantly, the boys also won the Junior Champion Eight of New South Wales. Not only were they pitted against all the other schools, but also against clubs like Sydney, Mosman, Haberfield and Drummoyne etc. They then went on to represent New South Wales at Lake Ballarat in Victoria against the rest of Australia, coming in a very creditable second.

I have always found sport to be a very rewarding and beneficial pastime. In fact, I have often thought that in business, as in other aspects of life, some of the easiest people to deal with are those who have a close involvement in sporting activities. Invariably they have learnt what sportsmanship is all about, win, lose or draw. And if they did lose, they did so with a smile, still able to say, "Well look, sorry I didn't give you a very good run for your money, but I'll try better next time". It is a valuable asset to have, no matter what walk of life you come from.

In 1948 Margaret and I and our young daughter Vivienne, who was only a baby then, went by ship to South Africa and took a flat in Johannesburg. We planned to stay for about six months, combining business with a bit of sightseeing. During that period I spent most of my time at the offices of our agents, McAdam & Bower, sorting and grading diamonds for use in Sydney. It also gave me an opportunity to get out and see some of the diggings with them and visit some of our friends in the diamond cutting business.

There were two factories in particular that I loved to visit. One was the Van Zwam's diamond cutting factory, run by Theo Van Zwam and his brother Joe. Theo's son, Hugo, was an electronics wizard and responsible for some incredible improvements in the diamond cutting industry. At that time, when computer technology was really in its infancy, the Van Zwams were among the first in the world to put all their wages, stock, debtors and other financial records on computer. They were well known to my family in Antwerp and had left there to set up in South Africa in 1929, around the same time my parents were also contemplating a move to South Africa.

My other favourite haunt was a diamond cutting factory run by the Abrams brothers under the name of Liberty Diamond Cutting Works. The Abrams were very knowledgeable and exceptionally good diamond cutters. I remember when my father used to go to Johannesburg, he would sometimes spend hours with them, examining some particular stone they had on hand at the time, deciding what techniques should be used and how it should be cut. At the time I visited them, they had a very large stone which they had bought from the Syndicate at Kimberley. Diamond cutters in South Africa were fortunate enough to be able to get their rough diamonds direct from De Beers at Kimberley. This particular stone was a perfect octahedron. It weighed 165 carats, but was yellow in colour, and not a very good yellow at that. A yellow diamond, if it is a good canary colour, is very desirable. The main problem with this stone was to decide what to do with it, in view of its enormous size. If you saw a stone weighing 165 carats, allowing for the fact that you will lose about 2½% in the sawing, you will finish up with two stones of about 80 carats each. If these in turn are cut into brilliants, you will end up with a pair about 40 carats each. But who on earth would want to buy one or two 40 carat stones of that particularly undesirable yellow? So they made the very wise decision to saw off the six points of the octahedron, yielding six individual stones weighing approximately 20 carats each. When cut, these would each produce about 10 carat stones. There was an added bonus too, since the colour of the sawn pieces improved out of sight and with cutting would become very saleable commercial whites. There was a ready market in the United States at the time for stones of that size and quality. The cube which was left over after the six points of the octahedron had been sawn off was a very dark yellow brown. It could either have been split or sawn and used for a number of different industrial purposes as in lathe or boring tools, optical drills, thread formers or wire-drawing dies.

Our agent, McAdam & Bower, had been buying quite a lot of diamonds for us from Dr. John Williamson in Tanzania. I was very keen to meet him and hoped to get the opportunity of seeing his diamond mine. As things turned out, our meeting did not eventuate on that particular trip. Nevertheless, it was a good excuse to set out on an extensive tour along the east African coast. I journeyed by ship to Lourenco Marquez, Zanzibar, Dar es Salaam and Mombasa. I had loaded the car on board in Durban and on reaching Mombasa did the rest of the journey to Nairobi by road. I encountered a lot of animals along the way, including a pride of lions which sat impassively in the centre of a dry creek bed, blocking the road and resolutely ignoring my earnest honkings to move on. I got a bad fright in one place where I stopped to photograph a magnificent pool carpeted with bright water lilies. I got out of the car and was just positioning myself for the shot when I looked down and stopped in my tracks. One more step and I would have trodden on a green mamba, the most deadly snake in Africa. At that moment I was only fifteen minutes from eternity.

When we finally returned to Australia, we travelled again by ship, stopping at Djakarta and Singapore then on to Perth, completing the final leg home to Sydney by air. It had been a very long trip, all in all, particularly with our young daughter in tow. At nine months of age, Vivienne took her first steps on board ship as we sailed back to Australia, lurching around on deck for all the world like a drunken sailor.

Our second child, Bill, was born in Sydney not long after our return from South Africa and two years later a second son, John, was born.

In early years, as the children were growing up, we made our home at Wahroonga, not far from where my parents were living at the time. It was a wonderful environment for the children, with lots of friends around and open space for them to play in. We entertained frequently and lived very comfortably with a large home and garden, a pool and well-equipped workshop. The children attended local schools at first but later on the boys went off to board at Our Lady of the Sacred Heart at Bowral. On our week-end visits, we would be up at the crack of dawn so that Margaret could roast a couple of chickens for them. They usually wolfed them down without ceremony in the back of the family car while Margaret and I attended church. Later on, Bill and John were both accepted as pupils at Riverview, and Vivienne attended Kincoppal at Elizabeth Bay.

I'm glad to say both the boys went to university, Bill graduating in Mining Engineering and John in Commerce with a major in Marketing. University vacations always saw them off somewhere to get some work experience. Bill went off to South Africa and spent some time at the Diamond Research Laboratories and worked as a Mining Engineer for De Beers. Back in Australia, he got more experience with mining companies before joining the family business, Jules Joris Pty. Limited. Further studies both here and overseas have given him a good grounding in business and gemmology and he now holds a managerial position with the Retail Traders Association in Sydney.

John has also travelled extensively overseas, and set off on a year long working holiday in Europe and the East after graduation. He had also pursued the Joris interest in gem stones, doing further studies in that field. He now runs his own business as an importer and dealer in precious stones, dealing mainly in sapphires and rubies.

Both my sons have learnt the art of cutting and polishing gem stones. It is, after all, a Joris tradition. It fills me with great pride to think that the example I have tried to set, and the training I have given them in various fields, has been as important to them as the training I received from my own father, and he from his father before him.

CHAPTER SIX

The Diamond King

Our agents in Johannesburg, McAdam & Bower, had during the war years been buying diamonds from Tanganyika, from a man by the name of Doctor John Thorburn Williamson, and occasionally they used to send these to us in Australia. Both Dad and I were delighted with these diamonds, which were excellent for industrial purposes.

Later on, after the war, when we were cutting gem diamonds again, we discovered that these "new" diamonds from Tanzania improved remarkably in colour. In other words, if they were slightly off-white in the rough, when they were cut they turned into fine white.

Needless to say, when I was invited to go to South Africa to attend a Diamond Drilling Convention in Johannesburg in 1952, I took the opportunity of getting in touch with Dr. Williamson and was invited to his mine. So the diamond trade brought me into personal touch with the man who had flashed like a meteor, and with near shattering effect, into the diamond world.

The Williamson story is of world-wide interest. Dr. John Thorburn Williamson was a Canadian who graduated from McGill University in Canada. When the De Beers company at Kimberley needed a first-class geologist, they advertised world-wide. He applied and obtained the job. His work was mainly on blue ground, and the nature of varying diamond-bearing deposits. Later on they sent him to Rhodesia to work in the copper mines down there. But Williamson's heart was set on diamonds. Instead of prospecting for copper, he asked the company if he could prospect for diamonds in Tanganyika. He explained he had a strong hunch that there were diamonds there. A few diamonds had been found in Tanganyika before, just the odd occasional stone. His experience and his geological knowledge made him stake his future on there being diamonds found in the wild, unexploited country. But the company turned him down. "If you want to look for diamonds," they told him, "you had better go back to Kimberley". How they were to regret that costly refusal!

Williamson resigned from the company and started prospecting on his own. It was a hard, lonely life. The money he had saved he spent on equipment, on tucker and so forth. He had a little T-Model Ford, which was his working transport. Constantly travelling while prospecting, he eventually became so poor that he was forced to borrow money from a bank clerk in

Nairobi. In fact, he could not even pay for his food, which consisted virtually of flour, tea, sugar and a few odds and ends. Determinedly battling on, at last he became so desperate he found himself owing money for a bit of tucker to an Indian store-keeper game enough to assist him. Then in 1940 when he was camped at a place called Mwadui, about 80 miles south of Mwanza, on the southern tip of Lake Victoria, he woke up one morning after heavy rains, started scratching around and found a diamond. Recovering his breath, between hope and fear he started to look further, to find, in feverish amazement, that he had actually camped on a large blue ground pipe, which under the right conditions can prove a home of diamonds. So, he promptly pegged out a square mile to start off with, put his pegs down, put the flags up, and left a notice-board with his name and date and particulars. Then he made for Mwanza to register his claim. When he got down there, he registered a claim with an Indian mining warden, and was half way back when he was stricken with malaria. He was feeling utterly miserable, when a native boy who worked for him came running to the tent and said, "Boss, somebody has pegged out inside your claim". Williamson had half a bottle of scotch left, which he downed, then got up on his feet and staggered to his claim. When he reached the site, there was a tent sure enough, right in the middle of his claim.

An Italian prospector had stumbled on the same diamond pipe, and, seeing Williamson's pegs, put his pegs inside Williamson's and advanced the date on his notice board by a month. Williamson knew, of course, that all the Italian would have to do was slap a five pound note on the Indian mining warden's desk saying, "I was there first!" and the mining warden would say "Yes", and grant the registration. Thus, Williamson would have had no claim in a court of law. There was only one thing to do. The foreigner was enjoying his afternoon siesta. No doubt his dreams were raining diamonds. He woke up with the muzzle of a gun poked right at his head, a wild feverish face glaring down at him. "I was here first!" hissed the wild man. "Now get!" The intruder got. Thus was found a great diamond mine, which startled not only De Beers, but the whole mining world.

Williamson immediately began producing diamonds in a quantity that increased amazingly after he sold a few to pay for labour and equipment. His second greatest thrill or pleasure, though, was settling up handsomely all those little, but so important, debts he had incurred during his battling days.

Resisting all attempts to buy him out, he would send diamonds from his mine down to the Standard Bank in Johannesburg. The bank would then get in touch with the leading diamond dealers, including McAdam & Bower, our agents. Thus, a healthy little Australian industry began buying diamonds produced by the struggles of a lone prospector in darkest Africa. Australian industry must have bought very many diamonds from Tanganyika. I sometimes wonder what will be the final result of the terrible upheavals over there, and I'm glad that Australia now has her own diamond fields.

I knew that a lively town had sprung up at Mwadui, started by a penniless prospector, but you can imagine my surprise when I landed in Nairobi in 1952 to find a De Havilland Dove (Dr. Williamson's private plane) waiting to pick me up and fly me to the mine. I was housed in the VIP guest house.

In those twelve years, from 1940 to 1952, the mine had developed enormously, now employing 2,000 natives and about 200 whites and about 80-90 Indians. Williamson now had two De Havilland Doves, a DC-3 and two Chipmunks. The Chipmunks were for any Europeans working at the mine who wanted to learn to fly. All they had to do was pay for the petrol.

Although initially Williamson was selling his diamonds on the open market in Johannesburg, eventually the Central Selling Organisation (CSO) made an agreement with him — a 5-year agreement — to sell his total production to them. This he did, and from his point of view it was quite a good set-up. Whatever he produced they would take from him, and pay cash for it at the end of the month. This allowed him to develop his mine and buy all the machines, including trucks, generating plants, housing, and everything else that is required on a mine. Williamson was of the opinion, however, that his diamonds were worth about 10% more. He was previously getting from De Beers ten pounds sterling per carat, "run of the mine" : that is, from the ordinary crushing boart through industrials to gem diamonds up to 14 carats. Anything above 14 carats was negotiated at a special price.

He had had a row with the CSO and he asked me what I thought he should do. He refused to sell to them, and he had a rather large parcel of diamonds on hand, worth 175,000 sterling which he had sent to the Standard Bank in London. However, he complained to me that nobody in London even looked at his parcel of diamonds. I realised why. He might have been a determined prospector and a marvellous geologist, but he certainly was no businessman — and diamonds are big business! I knew that in London the CSO sell their diamonds to a select number of buyers who have their "sights". In other words, people in the diamond trade get a "sight" from the CSO which they take up every five weeks. Most of them had been complaining to the CSO that they couldn't get enough and wanted a bigger allocation. They were rather surprised when they did get larger allocations, so much so that they had lots of diamonds but no money. It was then that Williamson's parcel suddenly appeared on the market. The diamond business is a cash business. In other words, if you haven't got the money, you just don't buy. You must take the diamonds with one hand and hand over the cheque or cash with the other. The CSO had also let it be known among those who had a "sight" from them that if they looked at Williamson's diamonds or made an offer on them, they would not in the future be very well received. So nobody was game enough to look at the parcel. There were not too many with the cash to pay for this particular parcel in any case. So the parcel just lay there.

Williamson was on top of the world, or seemed to be. He had to fight thoughout those battling, penniless days that led him to ownership of one of the richest diamond mines yet found on earth. He had another fight on hand now. He didn't fully understand it, and did not know what to do about it. Meanwhile, his rich, beautiful parcel of diamonds just lay idly in London. The proceeds were really needed to finance further operations, but foolishly he courted financial disaster by maintaining production and, as he did so, his overdraft kept going up and up and up, to such an extent that, when I was staying at Mwadui with him, he owed the bank 2 million sterling. It was a hefty sum even for a Diamond King in those days. Just as he had worked alone as a prospector, so now in this teeming financial world, he still tried to handle everything himself without the benefit of expert guidance. The result was a debt of 2 million sterling and a parcel of unsaleable diamonds. It was cold comfort indeed. Even with his apparent wealth and all his aeroplanes, Williamson felt nearly as bad as he had when he could not pay the Indian store-keeper the few shillings he owed for his tucker.

Because he took a fancy to me, perhaps, or because he felt I was not a part of the African diamond world or the equally cold and distant London financial world, or perhaps because I was just a visitor from some distant country called Australia and he was depressed and lonely, whatever his reasons, one evening he asked my advice on what he should do. I asked permission to think about it for a day or so. Then I explained the position to him with regard to ''sights''. The smallest ''sight'' people got in those days was £7,500 sterling, which progressively increased by increments of the same amount to £15,000, £22,500, £30,000, and so on. Nowadays, ''sights'' are $60,000 every five weeks for smalls, that is, from 1 carat in the rough down to 15 per carat, and $150,000 every five weeks for stones of 1 carat and up. I went on to explain that there were a lot of people in the diamond business who were not fortunate enough to have a ''sight'' from the CSO and were consequently only able to buy from those who did. My suggestion to him was this:

1. He should open an office in the Bahamas where there is no income tax or probate. His production cost was about £5 per carat, and he was selling to the CSO for £10. He could possibly get £12 on the open market. If he opened an office in the Bahamas, he could ship his diamonds to his own office at his own price, say £5.10s per carat, leaving only 10s profit for tax in Tanganyika. The tax in Tanganyika at that time was about 60%. From the Bahamas he could then send his parcels to his own small offices or companies in London, Antwerp, Amsterdam and Zurich using a higher price and giving him his profit in the tax-free Bahamas.

In fact, it would have been feasible to take the matter even further. He could have sorted and valued the diamonds himself and then sold them as ''run of the mine'' lots. This would have meant that both gem and industrial diamonds were mixed together, giving the buyer a chance to buy direct from the mine and avoiding any impression that they had been culled over and over. Buyers who deal with the Central Selling Organisation always know that every single stone they get to look at has been carefully valued and its resale value established. Illusion, and the possibility of getting a profitable bargain in a parcel, is still very important to the keen diamond man, and usually he would prefer to grade the stones himself, to suit his own and his client's requirements.

As far as overseas offices were concerned, it may have been possible to deal with a bank in say New York, London, Antwerp, Amsterdam and Paris or Zurich, and make arrangements with them to provide a small office with suitable lighting for say one of two days a month, and the use of their vault for security, in return for placing all remittances through them.

As the Central Selling Organisation has only about 200 odd ''sight'' holders, there are thousand of dealers who would be delighted to deal direct with an actual producer, without the goods having been graded into a couple of thousand categories and presented like premasticated baby food. It would have been easy to let it be known in the trade that ''sights'' of up to say £5000 in value, were to be offered on certain days in those cities. They could

have been made available to the small manufacturers and others not fortunate enough to have an allocation from the Syndicate. This would be likely to secure him a price of about £12 per carat.
2. He should then, with the profits made in the Bahamas, lend money to his own organisation in Tanganyika and pay say 7% interest on this money, which would be deductible from tax in Tanganyika.

I asked him how much he spent on insurance premiums, knowing that it must have been very expensive because of the number of employees and equipment he had. I then suggested that once he had accumulated sufficient funds in the Bahamas, he could look around and employ a good, competent insurance man and set up an insurance office in the Bahamas. Then, instead of paying premiums to London, all premiums could in future be paid to his own insurance company. The company could also act as an investment company for him so that all this eggs would not be in the one basket.
3. My next suggestion was that, although Barclays Bank operated a small office near the mine for the convenience of the mine and its employees, he could possibly start his own banking house at Mwadui, employing a competent banker and paying a small interest on current accounts. Most of his employees would probably bank with him anyway and the bank could give them a small 2½-3% interest on their current accounts. From the bank he could borrow money to further develop his diamond mine.
4. With the larger diamonds he produced, those over say 14 carats, (and he did produce quite a few large stones), the best gem-quality ones could have been cut in his own small cutting factory say in Geneva or Zurich, and sold to some of the wealthy oil people who came across regularly from the Middle East to spend holidays in Switzerland.

Williamson thought the whole scheme was an excellent one, but I think the sands had run out as far as he was concerned. The bank started putting the screws on him and wanted their 2 million overdraft drastically reduced. So he was compelled to go back to the CSO, eat humble pie and sell his diamonds to them. In the end this was probably the best thing for him because he was a wonderful geologist but a poor businessman.

It was later, when he paid his visit to Australia and stayed with us a couple of times, that he asked me to come and work for him at the mine, to name my own price, and to look after the gem and industrial side for him. But poor Williamson had cancer of the throat. It was written all over him that he would not last very long. The position would have been most interesting, but my job was in our own Australian diamond trade. Also, after all his struggles, Williamson would not have lasted more than 12 months and the business would probably be left to his brother and sisters. This happened when he died six months later. They knew nothing about diamonds, mining or geology, and eventually had to sell the mine to the Tanganyikan Government and De Beers company for 10 million sterling. It was dirt cheap. That mine was worth every penny of 50 million. The estate owed 2½ million to the Tanganyikan Government for probate. So the government paid 2½ million and the De Beers Company 5 million for a 50/50 share of the mine. Of that 7½ million his brother Percy and his two sisters got 2½ million sterling each. I realised that this would happen, and that whatever I could have done for Williamson would have been against the CSO, and possibly they would not have liked me for it.

Jules Joris, Dr. Williamson and Albert Joris on Sir William Walkley's boat on Sydney Harbour.

But I often recall the two visits I did make to Williamson's mine. He showed me some really fabulous diamonds, some of them green, some of them blue, some of them orange in colour, and he showed me a replica of the 54½ carat rose coloured diamond he presented to Queen Elizabeth, who was then Princess Elizabeth. He had it cut, and in its cut state it weighed 26 carats. It was the most beautiful pink diamond I have ever seen. Williamson had it mounted in the form of a sunflower with fine white carat stones surrounding it, with baguette shaped diamonds down the stem, and trapeze-cut diamonds at the bottom, then more baguette-cut diamonds forming a branch, and a large marquis diamond forming a leaf. Set in pure platinum, it was, without doubt, one of the finest pieces of jewellery I have ever seen.

Williamson died in 1958 at the age of 51. He was one of the finest men I have ever met, and one of the greatest geologist. One of the highlights of Williamson's last visit to Australia was his trip around Sydney Harbour with our very good friend Bill Walkley (the late Sir William Walkley) of oil fame. Sailing on Walkley's Impala, he was delighted with the marvellous sailing boats and our beautiful harbour. He was so impressed, that on the large dam he had for his water supply in Africa, he formed a yachting club, bought boats for his employees and also ordered quite a large motor cruiser which was dispatched from England in bits and pieces, assembled at Mwadui, and put on the lake. Unfortunately, he was never able to use the boat. He returned from Australia to become desperately ill, and died shortly afterwards.

After my first trip to Mwadui in 1952, I was flown back to Nairobi and from there caught a plane to Johannesburg to attend the Diamond Drilling Convention. I arrived at about 2 o'clock in the afternoon and stayed at the Carlton Hotel where there was a cocktail party on at 5 o'clock to welcome the guests. Sir Ernest Oppenheimer was the host to the 300-odd people who attended the cocktail party.

In diamond circles, this was really a world meeting, and I congratulated myself, a representative from the young Australian industrial diamond trade, on being given the opportunity to mix with these world heads to whom Australia was but a vague name on a map.

As I walked into the ballroom, Oppenheimer, acknowledged giant of the diamond world, came straight over to me, past all the large diamond buyers in the world and said, "Hello, Albert, how are you? How's Dr Williamson?" I found myself thinking, "You cunning old so-and-so! How did you know I had been to Tanganyika?" I never told anybody I got off the plane at Nairobi and spent a few days with Williamson at his mine. Sir Ernest was a fine, keen businessman who kept his ear to the ground. He was a man with a fine brain too. After Cecil Rhodes, he was the acknowledged father of the diamond industry in the present century.

Sir Ernest Oppenheimer died in 1957, and his place was taken by his son Harry Oppenheimer who recently retired as head of the Anglo-American Corporation. His business interests included the De Beers Company, Boart & Hard Metals, Diamond Abrasives and other companies. It would take a book to name all the companies and subsidiary companies which were ruled over by Harry Oppenheimer.

I learned a lot from my trip to the Diamond Drilling Convention, which took me not only to Johannesburg, but also to Kimberley and the fabulously wealthy Welkom Gold Mines in the Orange Free State. My trips to South Africa helped me to pick up knowledge that was to help later in our own Australian Mining Industry.

After the war, my father and I had established a company called Diamond & Boart Products Pty. Limited to continue the development and manufacture of industrial diamond tools. The war years had put an end to all work on gem diamonds for the jewellery trade and now, at last, Dad was able to again devote his energies to that side of the business. Gem diamonds would always be his first love and he gladly resumed his position as head of Jules Joris Pty. Limited. In those years when I was busy trying to get Diamond & Boart Products off the ground, it was my father's great enthusiasm for gem diamonds that more than once created a few headaches for me. He could not help but be fascinated with the task of solving the mysteries locked deep within the natural rough gem diamond. It presented a challenge that he must solve to his loss or gain and he loved nothing better than to meet and talk shop with people within the gem trade the world over. With great enthusiasm my father would seize upon the excuse of a trip to South Africa — well shephered by my mother, thank goodness. For Dad could never resist large rough diamonds, and a parcel of diamonds in the rough runs into money, believe me, even without the gamble of how they may turn out when cut. Not every diamond in the rough will cut into a gem stone, and every single one holds its own puzzling secrets.

I would be on tenterhooks, until the inevitable cable arrived, "Remit so many thousand pounds for a lovely parcel". Fortunately, he invariably bought well. When the diamonds were too big for our local market, I had to find an export market, generally in Singapore, Hong Kong or Beirut.

I remember one time he bought a perfect octahedron, slightly off-colour and weighing just over 50 carats. We sawed it in half to make a pair, each weighing just over 12 carats when cut and finished. Quite frankly, I didn't know how I could get rid of stones that size in Australia. Who wants two 12 carat stones as earrings, for argument sake? Through a friend of mine I heard that the Sheik of Kuwait had a passion for large diamonds. It was said that he wouldn't look at anything less than 9 carats and, for some reason or other they needn't be absolutely white. At least it was one little point in our favour. Eventually, contact was made through a friend of mine in Beirut, and sure enough, the Sheik bought both of them. It was a very happy release as far as I was concerned.

I remember one other time he bought a stone weighing 127 carats. I got strict instructions that I was to leave the stone until he arrived back in Australia. I was not to touch it or do anything to it at all. When I saw the stone I soon realised why. It was an unusual, fairly elongated type of diamond, and he had made up his mind when he was looking at it in Johannesburg what he would do with it. Upon his return to Sydney, he started splitting the diamond into various sections. I've forgotten the number of stones he split it into, but when he had finished, he stuck it all back together again with glue. The job he had done was really fantastic. The largest diamond weighed just on 18 carats when cut and there were a couple of 5 caraters as well as a number of smaller stones ranging between 2 and 4 carats. I sold all of these in Antwerp, except for the 18 carater, which we retained for a number of years before eventually selling it overseas.

On the subject of splitting diamonds, I might add that the largest diamond I have ever split was a 56 carat stone in Bombay in India. My father and I happened to be in India one time when we were asked by a mutual friend to have a look at a large diamond. My father expressed the opinion that the stone had to be split, and indicated where and why. They asked my father to split it for them on the pretext that they could not do it themselves without offending against their religion. It seemed a bit far-fetched and I have no doubt it was more a question of wanting someone of Dad's reputation to do the job for them.

So, with some prompting, my father started making the kerf or groove on the stone where it was to be split. He had an audience of several excited Indians who, seemingly unable to contain themselves, maintained a continuous, noisy commentary among themselves in their own language. It created such a distraction that Dad found it impossible to concentrate and, eventually losing his temper, he threw the diamond down and said, "Well, if this is how you want it, split the so-and-so stone yourself". Splitting a large diamond, particularly a good quality stone like that one, does take a tremendous amount of concentration, and you simply can't afford to have people moving around and talking all the time when you are trying to do this type of work. There is just too much involved. However, being a fair bit younger and more used to having noises around me, I took up the stone and continued the kerf where Dad had left off, successfully splitting it to the satisfaction of the people involved.

My father and I had our suspicions about how an uncut stone of that size came to be in India in the first place. It had all the hallmarks of a diamond from Dr. Williamson's mine in Tanganyika, and we had little doubt that it had been stolen. As a matter of fact, I had heard whispers of another diamond, a huge 350 carat stone, that had been stolen from Dr. Williamson and finished up in Bombay.

Dr Williamson himself later told me that he employed quite a number of Indians, and it often happened that when a decent-sized stone was found by one of them, they would conceal it and smuggle it from the mine. Wrapped in cloth or something similiar it would be heaved at night time over the eight-foot security fence which surrounded the Williamson mine and wait to be collected by somebody on the other side. There must have been quite a good understanding between these diamond thieves. If the police or security people picked up somebody with a diamond on him which they suspected of being stolen, you could lay odds on the fact that he would never reveal who his accomplices were, or who else might be involved in their syndicate. And he would go to jail confident that the others would look after his wife and children and ensure that they weren't short of anything until he was freed again.

Williamson was not a spiteful type of man at all. He knew very well that this sort of thing went on at the mine and in most cases he didn't even inform the police. To a certain extent, it was a case of bowing to the inevitable and, within reason, he was prepared to turn a blind eye to it. He was the type of man who didn't carry grudges and I admired him very much for that special kind of wisdom.

As for the 56 carat diamond we saw in Bombay, as I said earlier, both my father and I suspected it had come from Tanganyika. But the question of proof was another matter. Diamonds, unfortunately, do not carry a convenient label saying something like, "Produce of Tanganyika". They do not bear identification marks, except for people like my father and me and other diamond experts. There was little we could do about our suspicions in the circumstances. Nobody would have listened anyway.

It was not only my father's occasional buying sprees in South Africa that presented us with some business headaches. I remember one occasion when we had accumulated such a large number of industrial diamonds that we were faced with some liquidity problems. For some time after the war we had been getting substantial parcels of industrial diamonds through our agents direct from Johannesburg. We were able to use most of them but certain parcels were unsuitable for the type of work we were doing at the time and there was no market here in Australia for them. They had become a bit of an embarrassment to us with so much of our money tied up in them.

So one day my father said, "Albert, you had better go to London and see if you can get rid of those diamonds and send the money back here as soon as possible". That is exactly what I did.

Through a mutual friend, I made contact with a man called Marcel Mendelson. His family originally came from Antwerp, but he settled in London and become well-known in the diamond trade. At one time, Marcel did a fair amount of business in cut diamonds with Indian diamond merchants. In India, bargaining is an accepted part of business life, and an Indian buyer likes nothing better than the prospect of being able to get a parcel of diamonds or even a single stone at below the asking price. One had to expect, therefore, that negotiations would be fairly spirited. But Marcel used to be known in Bombay as, "One Price Mendelson". When he put a price on his diamonds he wouldn't budge from it for even one cent. In London, Marcel had an office in Hatton Gardens, in a building full of diamond people. It had a strongroom in the basement where the goods could be locked up for the night.

The diamonds I took to London were surplus to our requirements and were somewhat of a mixed bag, varying considerably in size and quality. Marcel contacted only those buyers he knew would be interested in industrial diamonds of that sort, and arranged a series of appointments for them. On each occasion he sat behind his desk with the prospective buyer in front of him while I took a position discreetly to one side and slightly behind the buyer's chair. I always kept a copy of the Times held at eye level, feigning a total lack of interest in the proceedings while Marcel, in Yiddish, conducted the business side of things, discussing the quality of the stones and so on with his client. Invariably I would be introduced as a friend from Australia, no name mentioned and my presence would be soon forgotten as the bargaining got underway. The buyers did not know, and Marcel certainly did not let on to them that I could understand every second word of Yiddish. With my Belgian background and fluency in both Flemish and German, I was able to get the general drift of the conversation and monitor the progress of negotiations.

As they bargained, I sat there mutely, and occasionally Marcel would glance over to me to see my reaction. If I shook my head very slightly, it meant that I wasn't satisfied with the price, and they would haggle some more. When eventually they had reached a price which was acceptable, a slight nod of my head would indicate to Marcel that he could close the deal, whereupon they would say "Mazlam Broka". After the deal was completed, the purchaser would pay in five pound notes. Five pound notes in those days were sheets of white paper of enormous size, known everywhere as "blankets". So bulky were they, I had to go out and buy a small briefcase, somewhat like a flat music case, in order to get the money safely to the bank. With anything from three to ten thousand pounds worth of "blankets" in my little case, I would make my way, either by tube or bus, to the Bank of New South Wales in Threadneedle Street, London. There the unfortunate teller would laboriously count it all out and double check it, before issuing his receipt. With two hundred five pound notes in a thousand pounds, it became quite an ordeal for the teller to have to stand there and count out several thousand pounds worth. On about the third occasion, the teller spotted me on my way into the bank, and I was greeted none to enthusiastically with a, ''Christ, not you again!''

Anyway, it didn't take long to liquidate the surplus goods, much to my father's relief. He was a man who simply did not believe in bank overdrafts. He would rather go without than be caught owing money to anyone. It was an attitude that was deeply ingrained in his nature and was a hangover from his days in business in Antwerp. There, the time honoured traditions of the diamond trade dictated that no matter how tempted you might be by a wonderful parcel of stones, if you didn't have the money, you didn't buy. It was as simple as that.

CHAPTER SEVEN

Indian Adventures

In a busy life, particularly when you travel around a lot, you will often get to meet interesting and fascinating people, people who you remember distinctly for a long time afterwards. One such person for me was a man by the name of Hassam Javeri. I first met Hassam in Johannesburg through dealings with our agents, McAdam & Bower.

Hassam was a very tall, very good-looking man from northern India. He dressed immaculately in suits from Savile Row and spoke fluent English, French and German. But, being an Indian, unfortunately, in those days, I could not take him out anywhere for lunch or dinner, so we just met in McAdam's office and talked about diamonds and various other things. We formed a quite close friendship and over the years I saw him regularly on my trips to London or Europe or India. He and his wife, Shirin, had a house in London and an apartment in Antwerp but most of their time was spent in Bombay, where they graciously accommodated me on one of my visits there. On that trip, in fact, Hassam was kind enough to accompany me on a long trip through most of India.

Hassam was the personal jeweller of the old Aga Khan, whose full name or title is Sultan Sir Mohammed Shah, head of the Ismailian Mohammedes. He has a religious following running into several million people, a number of whom live in East Africa. It was there at one stage that the Aga Khan is reputed to have been weighed in gold and later in diamonds.

Listening to some of the stories that Hassam told me about him, I think the Aga Khan must have been a very remarkable man. He employed two male secretaries who apparently he kept flat out with a constant stream of telegrams, cables, telexes and so on as he conducted his very extensive business empire. Rarely did he write letters. He had banking accounts all over Europe in all the capital cities, and he used to get banking and financial information on a daily basis from New York, Chicago, London, Paris, Madrid, Berlin, Brussels, and so constantly monitoring the latest developments in international financial markets.

Hassam had a very extensive knowledge when it came to diamonds, but he was also very well informed about other precious stones like rubies, emeralds and sapphires. He once told me of how the Aga Khan, wanting a special little gift for the Begum, had sent him in search of a perfect five carat diamond. So Hassam travelled from London to Paris to Brussels and

Antwerp, to Amsterdam, Zurich and Geneva, and one or two other places I have forgotten, in order to find the perfect five carat stone. When eventually he did find such a stone, he stayed and argued and bargained with the people who owned it, until eventually he bought it at an absolutely rock-bottom price. Knowing the Indian gift for turning a profit, I asked Hassam how much he had made on the deal. He said, "Nothing, not a penny". I said, "You've got to be joking. I have never met an Indian yet who would buy something on somebody's behalf and let him have it at absolutely cost price, travelling around Europe is expensive, staying at hotels and having meals and so forth is expensive. Do you mean to tell me that all this came out of your own pocket, and that the Aga didn't pay a cent towards any of it?" He said, "That's right". So I said, "You might be religious, and I know that you look upon the Aga Khan in the same way as Catholics would look upon the Pope, but all the same, I find it a bit much to believe that you would go to that much trouble and expense for nothing".

But Hassam had his head screwed on the right way. He told me the Aga Khan had a lot of followers in Eastern Africa. Some of them were very wealthy businessmen who had been in textiles, owned shops and all sorts of things. When they saw the writing on the wall and the inevitability of being turfed out of Kenya, Tanganyika, and places like that, they wrote to the Aga Khan, seeking his advice. They wanted to know how to go about getting their money out of the country without attracting too much attention. They couldn't put a house, a store or a block of land in their pocket, and they couldn't buy up gold because, with its great weight and bulk, they would not be able to move it out easily. So the Aga's advice was to buy diamonds. "See Hassam Javeri," he would say. "He's the man, he's honest, knows the business, knows diamonds, he knows where to buy them, so get them from him." And that is where Hassam made his cop.

Hassam was always keen for me to go into business with him and to set up a diamond tool factory in India. Had we done so, it would have been the first such factory in India. The big problem was that I had been to India on several occasions, and every time I had come back with dysentery. In fact, the last time I was there I contracted amoebic dysentery, which took me eighteen months to shake off. I think it probable that the episode has caused me some permanent liver damage.

He was also very keen for me to set up a casting factory. It was not to be an ordinary foundry, but one employing the lost wax process. The lost wax process had been used extensively for thousands of years, not only by jewellers, but also by manufacturers of precision engineering parts, such as in the manufacture of the small cog wheels used in sewing machines for instance. The broad principle of it is that a master craftsman will manufacture, for argument sake, a ring, brooch or other piece of jewellery in silver or other metal, and from this master design a rubber mould is made, producing a cavity of the exact shape of the article to be cast. Then molten wax is injected into that mould under pressure, and once it has set the mould is taken apart leaving a wax ring or brooch or whatever. The same mould is used over and over again until the desired number of articles has been made, all in wax.

Then what is known as a tree is produced. That is to say, there is a wax stem which looks like a tree in a way, very wide at the bottom, gradually tapering up to the top with branches coming out of it, and on each branch one of the wax rings or whatever the casting may be, is attached. Then the whole thing is dipped into a silicon solution which sets and hardens, leaving a solid tree.

This is then put into a low temperature furnace and gently heated until the wax melts and runs out into a container. Then, basically, you are left with a hollow tree trunk with hollow branches with 50, or perhaps 100 rings, brooches or whatever attached. Molten metal is then poured, and later the silicon is removed, and each part is detached from its branch and filed clean. And there you have it, 50 or 100 parts in a finished state. Such a factory would have been a first for India, but as I said before, I was not keen to go and live there for the considerable periods that such business undertakings would require.

Hassam was a man with a lot of novel ideas. I remember one time when I was in Europe, I looked for him in London, in Amsterdam and Paris, but each time he would be somewhere else. When I was in Amsterdam, he would be in London or Paris, and we seemed destined not to meet at all. But I finally caught up with him in Geneva in the hotel du Rhône. I was in the lift coming down from my room when the doors opened, and there he was. At once he said, "Albert, come upstairs quickly. I must show you something." So we went up to his room and he picked up a briefcase he had left there on the floor. I was absolutely staggered when he opened it up. I

Hassam and Shirin Javeri in Bombay.

had never seen anything like it in all my life. The briefcase was chock-a-block full of emerald jewellery.

He explained that the Nizam of Hyderabad had died, leaving various quantities of emeralds in a number of Swiss banks, and Hassam had been instructed to dispose of them. Apart from brooches and rings, the cache included some of the most beautiful necklaces of hand-carved emeralds I had ever seen. I asked Hassam where on earth he was going to sell them. They were undoubtedly worth a fortune.

He said, "I'm not going to sell them. I'll break them up and saw the beads in half and make two emerald cuts out of each stone." "You can't do that," I said, "it would be a crime". He said, "Well you tell me where I can sell them as they are, and I'll go and see the people straight away". He was right, of course. There was just no market for that kind of merchandise. The briefcase contained about two, to two and a half million pounds worth of gems. I was absolutely flabbergasted that he could quite nonchalantly leave it sitting there on the floor of his room while he went downstairs. "Well," he said, "nobody knows about it, except for you".

At one time, the maharajahs in India were like kings in their own domain. They lived, for the most part, in absolute luxury, immune from the burden of taxes. Once India was federated, however, they suddenly discovered that they had to pay taxes to the Indian government. It meant that a lot of their goodies, their antiques, their jewellery, their ivory and other valuables had to come to the surface in order to pay off their massive tax debts. It was then that Hassam had the bright idea of chartering a twin-engined aircraft, which he flew from one maharajah to another. He came back with 400 kilos of jewellery, gold, silver, precious stones, and anything of value. Apart from himself and the pilot and fuel 400 kilos was about all the aircraft could carry and still get of the ground.

On one of his visits to Australia, Hassam came to me for assistance with some business arrangements he was making on behalf of the young Nizam. The Nizam wanted to acquire property in Western Australia and settle there. Hassam was to make all the appropriate arrangements. In those days, you had to get government permission to bring large sums of money into the country. I think there are still some government restrictions on foreigners wishing to buy Australian real estate. Anyway, at that time, Hassam wanted to get the proper clearances to bring in a quarter of a million dollars to buy a particular property with a fifteen mile beachfront south of Carnarvon in Western Australia. He also had instructions to find a 100 to 120 ft cruiser for the Nizam, capable of at least 18 knots. Fortunately, I was able to be of some assistance with both missions, and the young Nizam's requirements were satisfied.

At one time Hassam and I did a very extensive trip right through India. It made all the difference in the world being able to travel with someone who had an intimate knowledge of the country and the language. Not only that, Hassam's position and influence gave him entry to all sorts of places which, had I been on my own, would never had been open to me. I felt privileged indeed to have him as a travelling companion and guide and he took me to some wonderful temples, museums and other places of interest.

Perhaps the highlight of that tour was a visit we made to a palace in Hyderabad. It had been the home of a nawab, a man who, in our society, would probably have a social status equivalent to that of a duke. The nawab had been a bachelor all his life, had evidently enjoyed fabulous wealth, and had been a discerning collector of *objets d'art.*.

At his death, he left his palace and all his fabulous art treasures to the State. A French expert was called in to assess the estate and advise on the best means of exhibiting the enormous priceless paintings and artifacts housed in the nawab's palace. He recommended that a 150-room museum should be built to do the collection justice. The cost would have been prohibitive and the State could not afford such an expensive undertaking so, in the end, it was decided to preserve the palace, leaving the entire collection displayed just as the nawab had arranged it himself.

Hassam took me through this palace and I was absolutely staggered at the wealth it contained. There were things there I could not even have imagined had I not seen them for myself. Apart from paintings by some very famous artists, there were statues from Italy, France, Belgium, England and a host of other places in Europe. The rooms were enormous and several were devoted to some extraordinary items forming individual collections. There was a Gold Room for instance. Every article in that room was made of solid gold. It housed gold dinner sets, gold drinking vessels, gold cutlery, fantastic gold jewellery and other trinkets. Then there was the Silver Room, again containing nothing but pieces made from solid silver. There were silver strappings for elephants and horses, silver platters and goblets and all manner of silver ornaments and jewellery. There was also a Jade Room. One piece that particularly caught my eye was a small jug, much like a cream jug, which stood about 20 cm high. It had a stopper on a chain and a handle at one side. The whole thing had been hand-carved from a single piece of jade. Can you imagine the incredible skill involved in carving each perfect link in that jade chain, the stopper, the handle and jug, from a single slab of jade? The whole thing was quite exquisite.

Another extraordinary item that caught my attention was a white jacket, much like those worn by wealthy Indian men. This one came to just below waist level with long sleeves and a small stand-up collar. It buttoned right to the neck with a single row of seven buttons down the front. I believe the number seven has some religious significance in India. Then it had four buttons on each sleeve. The amazing thing about that jacket was that all seven buttons on the front were diamonds of approximately ten carats each, and each sleeve sparkled with four smaller diamonds of approximately five carats each.

I think I could have spent a week there without seeing it all. I wish I had had more time. It was perhaps the most remarkable display of wealth I have ever seen assembled in the one place.

So my travels through India with Hassam proved most entertaining. He made sure that I didn't eat or drink anything that had not been properly sterilised. All green vegetables, such as lettuce, for example, had to be dipped in a solution of condy's crystals rather than just washed in water, and all my drinking water had to be either double distilled or double boiled. During the time I was in Bombay, Hassam and Shirin insisted that I stay with them, and they looked after me very well.

How different it was to my first visit to India when I went through on my own. I entered at Trichinopoly in the south on a DC3 from Colombo. Only later did I find out from Hassam that DC3s were made in India under licence. I would not have felt so confident if I had known that at the time.

In any event, when I arrived at Trichinopoly, I had arranged to be met by representatives from the import/export company we had been dealing with from Australia. Their letterheads disclosed a rather impressive list of company executives, but on arrival I discovered that they were all members of the one family. When my plane touched down I was greeted by a sea of faces. Father, grandfather, uncles and cousins had all turned out to meet me. They had booked me into an old English Club and took me there by car.

When I saw the rather primitive hotel facilities available elsewhere in the town, with five or six to a room, I was grateful indeed for the presence of an English Club. Being the only guest at that time I was very well looked after. Later they called to take me to their home. On the way one of them whispered, "Would you like a drink?" I had not realised that this part of India was dry, as is most of India. They stopped at the entrance to a little lane and somebody stood look-out while another dashed in through one of the doorways emerging a short time later with a large parcel. I was told not to unwrap it until I got back to the English Club later that night. When I finally did open it, I found a bottle of scotch, carefully wrapped in layer upon layer of newspaper.

The family had recently acquired a new home, reached only by a short but confusing journey through the narrow streets and laneways. We drove until the car could not travel any further and covered the rest of the distance on foot. Then we came to a newish-looking house, with all the rooms built around a central courtyard. I was shown into the main room and offered a place at the table. There must have been about 20 people there, while out in the courtyard all the women of the household were gathered, peering around the edge of the doorway, trying to catch a glimpse of the stranger from Australia.

A meal was served with a large assortment of exotic dishes and I was asked to try this, and try that. The first thing I tried burnt my throat so badly I thought it must have been nitric acid or something similar. It was a meat dish coated in some reddish sauce which they seemed to lap up without ill effect. I told them as politely as I could that, having already eaten, I really wasn't very hungry but they pressed me to try another dish, greenish in colour, which was put before me. Not wanting to offend their hospitality, I took a mouthful. It was unbelievable. My mouth felt as if it was on fire and I looked around for something to drink to cleanse my throat. I don't know how I managed to get through that meal. At the end of it I noticed that all my companions started burping and I soon realised that it was the excepted thing. A loud burp showed your appreciation to the host for the wonderful food. So I took several deep breaths and eventually managed a small burp, hoping that at least I had observed the proprieties.

I was shown through several factories where semi-precious stones were cut. They used quite a lot of diamond dust in cutting the stones and much of it was made from the large quantities of the crushing boart we supplied. Crushing boart is the cheapest of the cheap. As far as diamonds go, it is just the rubbish left after industrial diamonds are sorted out, and is good for nothing except crushing into diamond dust and diamond powder for abrasive purposes.

All the factories were using synthetic sapphire and ruby boules, manufactured mainly in Switzerland. These man-made stones are shaped more or less like a bottle, only smaller, about an inch in length at the most, and are sliced into pieces and then facetted to make either brilliant, square or rectangular cut stones. These are used for making cheap costume jewellery. The cutting and facetting of the stones requires diamond dust. But the Indian factory owners would not buy diamond dust from us or anybody else for that matter. They always had a feeling, I suppose, that they might be cheated and get cigar ash instead of the very fine, micron-sized diamond dust they needed. So what they did was buy cheap boart diamonds and crush them themselves. That way they knew that they would get 100% diamond dust without the risk of being cheated by European suppliers.

I think it took just on two days before I had a severe bout of dysentery, but I had so much business to get through I just had to carry on. I went south to Madras to see more of our clients and from there to Bombay. Bombay was then a major centre for processing precious stones and now there is a very big diamond-cutting industry there. From Bombay I went north to Jaipur, where again there is a very big market in gem stones. Wherever possible I stayed at government run guest houses which were reasonably clean and provided edible food. On my last visit to Jaipur, I stayed at one of the palaces owned by the maharajah. It was an enormous place and had been turned into a hotel. The rooms were large and comfortable but I found the food unbelievably bad.

Jaipur had quite a large industry in cutting precious stones but the techniques employed are terribly primitive by western standards. They use the same methods they have used for hundreds of years, usually without the benefit of even an electric motor. For the most part, the Indian cutters and polishers I saw worked at a simple wooden cylinder with a piece of string wrapped around it and a spindle running into a couple of wooden forks. At one end an abrasive wheel was pulled backwards and forwards with a piece of string, much like a yo-yo, but it bounced around so much I don't know how they managed to polish a facet with it. I did see one little sixth of a horse power electric motor, but I would not have been game to touch it in a fit. It had an exposed wire that was almost bare at one place, and on the spindle a short length of what looked like a broom stick had been mounted. It seemed to bounce uncontrollably all over the place. There was some diamond dust on the wheel and they were polishing stones on it, but it don't know how they managed. Of course, time and labour was not a problem there for the factory owners. Wages were extremely low, with the best tradesmen, working from sunrise to sundown, getting no more than $10 or $12 dollars a week.

At one point I believe a committee was set-up to look into ways of modernising the industry. It was decided to import modern gem polishing equipment from Japan and Europe but, unfortunately, when it arrived no one knew how to use it. To this day, I think that equipment is probably still sitting in a store somewhere gathering dust. It's a great pity that no thought was given to sending people off overseas to get the specialised training needed to operate the machinery. As a result, the produce they turned out was extremely poor and unsuitable for use in the West. What tends to happen is that European buyers, especially the French and Swiss, go to India and select cut stones of suitable colour and quality. These are then taken back to be recut or refacetted in Europe in a way that will make them acceptable for the European or American market.

It is the same with diamonds cut in Bombay. The cutters are quite skilled in recognising the diamond grain and they make sure the disc is running at right angles to the grain when they put a facet on. The end produce is poorly cut, however, and is really only suitable for some of the very cheap jewellery made in the East in places like Bangkok, Hong Kong and Singapore. I suppose a lot of people outside the diamond trade wonder how diamond cutters in Antwerp, Amsterdam, New York or France can possibly make a living where their wages are a least 30 times higher than those paid in places like Bombay, for instance. The diamond industry is highly labour intensive and labour costs can be significant, particularly where smaller stones are concerned. The only reason European cutters are still able to compete is that they are absolutely top-class tradesmen and their cutting techniques are unsurpassed. No doubt diamond cutting in places such as Bombay will improve but I think it will take some time and no doubt we will see labour costs increased there as well. Be that as it may, my many visits to India always proved interesting and informative, and I am glad of the opportunities I have had to visit there.

CHAPTER EIGHT

The Career of a Young Apprentice

Maurice and Ronnie De Stoop, who came to us before the war, had started on their own. But there were many others whom we trained. People like George and Max Buckley. Today, Max works in the United States and is very highly regarded in the industrial diamond tool field. One time junior golf champion of New South Wales, Max learned his trade from us, decided to go overseas, and asked me if I could get him a job en route. I was able to arrange a job for him in Bombay, and within two weeks he was in charge of a gem cutting factory employing forty people. Then he went to London and worked for an industrial diamond tool company at Basingstoke, outside London. From there he went to Windsor in Canada, and worked for an industrial diamond tool company just across the water from Detroit. Later on I set him up in an office in Vancouver to cut gem diamonds, but unfortunately that did not last too long and he decided he wanted to go back to Windsor. I think a young French/Canadian lass might have had something to do with his decision at the time.

There were many others too. People like Barry Stoddard, who until recently was with Diamond & Boart, Harry Goddard, John Halloran, Bob Sayer, Tony Maynard and Osie Wagner, just to name a few. All became top-class tradesmen. I really believe the thorough training we were able to give them right through from start to finish, would not have been available to them anywhere else in the world.

In the early 1950s we took on a young apprentice at Diamond & Boart Products by the name of Tom Bain. At his mother's urging, Tom came to us straight from school. Mrs. Bain had made up her mind that young Tom was not the scholastic type but she believed he had some talent working with his hands and persuaded us to give him a chance. It has proved to be a long and valuable association.

Tom started in the diamond bit section which meant turning graphite, drilling holes in it and dimpling it. It is a very dirty job really because graphite dust flies all over the place and by the end of the day you finish up looking like a coalminer. But Tom seemed to revel in it. As he gained more experience, he started to develop a keen interest in diamond drilling. One day he approached me to see if I would help him get a job as an assistant diamond driller with one of the mining companies. As it happened, I knew the head of Australian Drillers. He was a Canadian by the name of Larry McCabe and after some discussions he agreed to take Tom on and give him some training. Tom worked in various parts of Australia, eventually finishing up at Mt. Isa mines working on both surface and underground diamond drilling operations.

I remember that in those days we had a standing order from Mt. Isa for 500 XRN bits per month. XRN bits are blast hole bits which do not produce a core. They merely drill a hole of 1¼ inches in diameter through the rock. Because of the amount of business I did there I used to go to Mt. Isa on a fairly regular basis. In those days, the trip to Mt. Isa meant leaving Sydney late in the afternoon with an overnight stop in Brisbane and a dawn start to fly to Townsville the next morning. From there it was the milk run by DC3 to Charters Towers with numerous stops at small places along the way before eventually arriving at Mt. Isa late in the afternoon. Landing at the old Mt. Isa airport was always a hairy experience. It was a short field landing every time and the heat and somewhat rarefied atmosphere made putting the plane down an arduous task for the pilot. As a passenger I didn't realise that at the time and it was only later, when I took up flying myself, that I discovered the difficulties involved.

I usually stayed at the guest house at Mt. Isa, which was very comfortable and it used to amuse everyone no end when I put my pyjamas in the refrigerator during the day time so that they would be nice and cool to put on at night before I went to bed.

Mt. Isa was the kind of place where an impromptu party would be thrown at the drop of a hat. It didn't need much of an excuse. Either somebody would be leaving or somebody arriving, or somebody would be getting married or going away on holidays or just getting back from one. At night time in Mt. Isa, there was always some fun to be had.

As it happened, we often used to go underground during the night shift to test out new designs for diamond drill bits. On one particular occasion we had been to a party beforehand and were pretty merry. You might say we had had one too many. So there we were, down in the bowels of the earth, with the noise of the diamond drilling machines sounding like machine-gun fire in our ears. Then they started blasting at the other end of the shaft and the acrid smell of cordite started drifting through the tunnel to where we were working. By the time we came to the surface early next morning we were all suffering with what the mining fraternity would call a fracture headache. I had an absolute father of a hangover. It was extremely painful and probably the worst of my entire life. I still wince at the thought of it, even today.

After a party the night before, the flight back to Townsville used to be quite an ordeal. I remember one time being on a DC3 flying at about 2000 feet, when Tom Bain and I had to nurse the air hostess who was so sick from the bumps and turbulence that she could not even see the passengers, let alone look after them.

Tom eventually went overseas and spent some time in London. At my invitation, he met me in Paris one time for a weekend on the town. We decided to do all the night clubs at Montmarte and Montparnasse. From experience, I knew how easy it was to get stung in those places. If you sat at one of the tables they would invariably come around with a bottle of champagne and press you to buy it. It would cost the earth and you probably wouldn't drink it anyway. So we adopted the strategy of standing up at the bar, having a scotch and soda and watching the show. When it was finished we would move on to the next nightclub, unwilling to call it a night until exhaustion took over somewhere around six the next morning. By the end of the second night, Tom claimed he could see everything double.

He later went on to Canada and worked for Boyles Brothers in Vancouver. There he met and married a delightful English girl and I invited him back to Australia with his bride to take up a position with Diamond & Boart Products Pty. Limited.

By then the company was making more and more diamond drill bits for the mining industry. One of our largest clients, apart from Mt. Isa was the Zinc Corporation, now CRA. The man in charge of drilling was Eric Cottrell who knew his drilling diamonds backwards. I think, over the years, we must have made tens of thousand of diamond drill bits for that company, particularly in the years they were developing Bougainville. At that time, Frank Paholski was in charge of the Bougainville project in its development stages. We were making so many diamond drill bits for them we had to work Saturdays and Sundays in order to get them out on time.

One of the big breakthroughs in diamond drilling was the development of the wireline core barrel. In the conventional method of drilling there was a

Diamond drill bits used by C.R.A. at Bougainville.

diamond drill bit, behind it a reamer, then the core barrel and then finally the drill rod string. One would drill down until the core barrel filled up, being 10 feet in length, the whole string of drill rod would have to be heaved to the surface. When drilling was being done to a couple of thousand feet there was a tremendous waste of time pulling it all up so that the core could be removed from the core barrel. Then the whole lot would have to go down again.

With the wireline equipment, the whole lot stays down and only the inner tube is picked up and brought to the surface by steel cable, for the core to be removed. Whilst that is being removed, a second inner tube can be dropped down, and drilling can continue until at last the bit eventually becomes blunt. When that happens the whole string must be brought to the surface for a new bit to be fitted.

Needless to say, I was very anxious to introduce this new method into Australia. Knowing that the Longyear Company in Minneapolis were by far the most advanced in this technique, I contacted them and told them that I was very anxious to set up with them in Australia. Apart from core barrels,

Longyear NWL wireline 9" series core bits.

they also made diamond drilling machines, and all ancillary equipment necessary for diamond drilling. As it happened they supplied the equipment and we made the diamond drill bits, which we sold under the name of Longyear on an equal share venture.

Again, Tom Bain was asked to manage the Longyear company in Australia because he had, by this time, quite a lot of drilling experience not only in Australia but also in Canada. The Longyear company was closely connected with another American company, Christensen.

I first met George and Frank Christensen at the Diamond Drilling Convention in Johannesburg in 1952. They were two huge men, both all-American footballers, and when they came out of the army after the war they decided to set up a factory to manufacture oil well bits at Salt Lake City in Utah. Although sharing the same surname, they were not in fact related. Their partnership, however, proved to be an enormous success and they set up a number of overseas branches to manufacture oil drilling bits in South America and on the continent.

Coal drilling bits designed by the author using synthetic polycrystalline and natural diamonds.

Since oil had been found in Australia, and there were a number of companies actively drilling for oil, I thought the opportunity was right to start manufacturing these large diamond oil well bits here. I realised that apart from the knowledge and expertise that goes into making such bits, a vital factor for success would be the availability of servicemen and all the ancillary equipment. This all pointed to a fairly substantial investment on the part of anyone contemplating such a venture in Australia. So I decided to go to Salt Lake City to see Frank and talk to him about the possibility of setting up a 50/50 joint venture in Australia. It didn't take me very long to persuade Frank that this could be a worthwhile undertaking.

I arrived in Salt Lake City and checked into the hotel around lunch time and after putting my suitcases in my room went down to lunch. In the hotel restaurant the waitress came along and said, "What would you like to drink?" I asked for a dry Martini on the rocks, but was politely informed that, as Utah was a dry State, no alcohol was permitted. "You may have a Seven Up or a Coke," she said.

After lunch, I spent the afternoon at the Christensen factory. It was set up much along the lines of our own operation in Australia although on a much larger scale. Around 5 o'clock a number of the foremen started wandering into Frank's office and sat there looking at their watches. I was wondering what it was all about when suddenly the 5 o'clock whistle sounded. It was evidently the signal to open up the bar. Our came the bottles of scotch and everybody got stuck into the grog.

That night Frank and his wife took me to his club. It appears that all the hard liquor was stowed away in the members' lockers, and at dinner found its way to a safe position beneath the table. Glasses of scotch or bourbon kept appearing all night with not a bottle to be seen anywhere. After dinner, the bottles went back to the lockers from whence they came. When I got back into my room that night, I found a bottle of scotch and an ice bucket and bottles of soda water, compliments of Frank Christensen. For a dry State, I have never seen so much grog in all my life.

Following lengthy discussions we decided to go ahead with the joint venture and I returned to Australia. Frank had expected me to lose money in the first twelve months, break even in the second twelve month period and start showing a profit in the third year of operations. We had agreed that Christensen's would supply a manager cum technical adviser for the venture. The manager and his wife duly arrived and I set them up in a motel at North Sydney whilst we drove around looking for suitable accommodation. I assigned some of our best fitters and turners and other tradesmen to him so that the operation could get underway. In the meantime, I did my best to help him settle in by taking him to Brisbane, Melbourne, Adelaide and Perth, introducing him to all the people likely to be interested in oil drilling and oil exploration and helping him to establish useful contacts. As you know, in Australia, when you have been talking business all afternoon, the boys will usually head down to the local pub and have a few beers. Everybody is expected to take their turn and shout a round. It is an unwritten rule. However, our new American manager did not seem to appreciate this rule. He did not seem to be able to take the hint. Eventually I think people were beginning to wonder whether he had a death adder in his pocket. Finally I just had to say to him. "Look, it's your turn to shout."

Even this fell on deaf ears. He had previously made it clear to me by some comment or other that he had not come to Australia to make friends.

Soon after I got him established in a factory on Parramatta Road, Stanmore, we started production and in the first six months made $40,000 profit. Then he said they needed larger premises. I found a large factory for him at Dee Why and arranged the lease and moved a lot of equipment there, together with the men and some additional equipment I had purchased. As the factory was in the Sydney area, I was able to go round and see them two or three times a week to see how things were going. Eventually he again decided they needed more space. So we looked around and decided to relocate the entire operation to Adelaide. It had the advantage of being very central, particularly for that type of work, and offered easier access to Western Australia.

Sir Thomas Playford was at that time Premier of South Australia, and he was a man with a tremendous amount of get up and go. In fact, the Mines Department was one of his pet projects and he seemed to enjoy a lot of time on drilling holes thoughout his State. If his drillers came up with anything worthwile, he would contact some of the bigger firms overseas and say, "Well, we have got this and that. Are you interested in coming to South Australia?" In many cases they did. Also I soon discovered there would be some very attractive incentives when it came to tendering for diamond drill bits for anybody prepared to manufacture them in South Australia.

When I saw Sir Thomas I said to him, "Now you do a lot of diamond drilling here in South Australia. How would you like a factory to make bits here in Adelaide?" He was delighted. I told him all we needed was land and he asked me where I wanted it. I said, "Well as close as possible to the airport." My reasons for suggesting such a location were that oil bits are pretty expensive items and when people need them, they need them in a hurry. An oil rig costs tens of thousands of dollars in the field, and when they strike hard ground they must have diamond drill bits. When an order comes in for an oil well bit, you have to be prepared to start work on it straight away, whether it is morning, afternoon or night. And you must keep working on it, right through the night if necessary, to get it finished and on to a plane without delay. If there are no landing strips near the rig, then you must arrange to have a carrier with a jeep or a utility waiting on hand to pick up the bit and take it straight to the drill site.

Oil bits are not the sort of items you can manufacture and put on the shelf. Different conditions demand bits of different design and diamond sizes. That is why it is essential to have a serviceman on the drill site. Of course, his services are paid for very handsomely by the oil company, but it is his expertise that will determine the design of the bit that is required. He must calculate the number of revolutions it should be doing, the pressure on the bit, and the type and volume of the fluid to be used as a coolant.

The Department of Housing to which Sir Thomas sent me, found me a two acre site right next to the airport and not very far from Glenelg. It was a very convenient position because Glenelg had a number of motels and a golf course and was only four and half miles out of Adelaide itself. So I built a factory on the land and we moved all the equipment there during the Christmas break. We were back in production by about the first or second week of January. The wiring for every machine had been laid on, so the moment the machines arrived they were unloaded, coupled straight up and ready for production with the minimum of delay.

The American manager then started to order large quantities of core barrels, innertubes, core lifters, stabilisers and a hundred other items running into tens of thousand of dollars. We made this equipment available to our customers either for purchase or for hire.

When everything was organised I made arrangements for an official opening. A number of people from the Mines Department, including the then Minister for Mines, were invited to attend. When everyone was there, I got up and gave a little welcoming speech and then invited the Minister of Mines to say a few words. He looked a pretty rough sort of a diamond, who extracted a crumpled bit of paper from his pocket and started to read. I was somewhat surprised that he had found it necessary to have a prepared speech considering it was not a terribly formal occasion and just a few words from him would have been sufficient. Afterwards I saw some of the boys from the Mines Department and asked them about it. It seems they were worried that some female staff from the factory would be present along with visitors and they dared not trust him to make an off-the-cuff speech, knowing his propensity for colourful four-letter words.

Unfortunately I could not go down to the factory two or three times a week as I had been able to do in Sydney. Because of the distance, I could

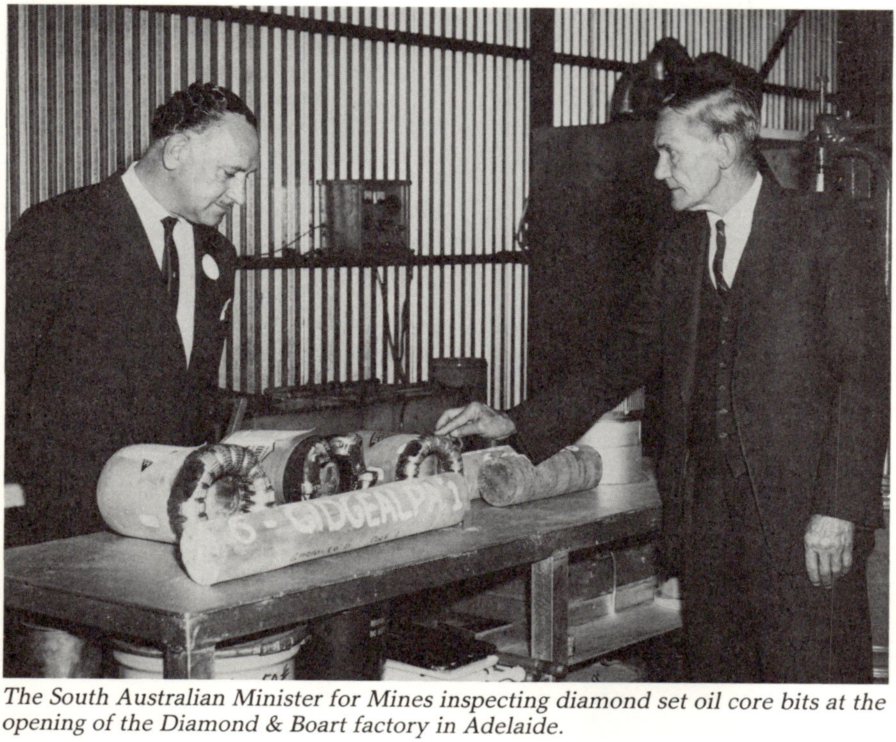

The South Australian Minister for Mines inspecting diamond set oil core bits at the opening of the Diamond & Boart factory in Adelaide.

only manage to go down there about once every four or five weeks. Before long, the factory started losing money. I was becoming very concerned and in despair rang Frank Christensen in Salt Lake City asking him to get rid of the American manager. I knew I could replace him with a local man at a third of the salary. But Frank would hear nothing of it. Eventually the postition was such that I had to get in touch with Frank again. I told him straight, "Either he goes or we split up the company". Ultimately we did agree to part company.

I suppose the 50/50 joint ventures with Longyear and Christensen's were two big mistakes I made. The marriage of a giant and a dwarf never works out because the giant, eventually, is going to come out on top. I have seen this pattern repeated several times since in Australia. Large American companies have come in here, used a local manufacturer to find out who is who, and what is what, and become familiar with the place. Once they know the layout and have all the right employees and contacts, the local man is punched out of the way.

So it was that after establishing both Christensen's and Longyear in Australia, I pulled out of both companies. Christensen's and Longyear eventually joined forces in Adelaide and built themselves a new factory and I carried on making drill bits for the mining industry in both the Adelaide and Sydney factories. From my point of view, it was a great relief.

Later Longyear became the property of the De Beers company, or rather Boart & Hard Metals. De Beers also made a bid to acquire Christensen's, I believe, but eventually it went to the Norton Company. Norton and De Beers are very closely related in any event, and this could have been a very convenient way of getting around American anti-trust laws.

It was a pity in a way that we had to part company. If it hadn't been for the enormous losses that the Adelaide factory was showing, I believe I could have raised the money to keep going. It would have been very interesting in view of the renewed activity that has taken place in Australia in the search for oil and gas. It would have been an even more viable proposition now that we are finding industrial diamonds in large quantities in Western Australia. Maybe there is room for it yet. I have had first-hand experience in setting up a factory like that, twice in Sydney and once in Adelaide. For my part, it wouldn't demand such a great effort if I was asked to be involved. It is certainly something to think about.

When we split up with Longyear, Tom Bain had the opportunity of either staying with Diamond & Boart, or going with Longyear as their Far Eastern representative. Tom chose Longyear, for which I don't blame him. After all he had been trained to make diamond drill bits, to handle and sell diamond drilling machines, pumps and all of the ancillary equipment that goes with them. He liked and understood the Far East, so he took the job.

I remember one incident he told me about. He had been trying to obtain a very large tender for diamond drilling equipment from a certain overseas Government controlled corporation. He had been seeing the Purchasing Officer there and got to know him fairly well. So one day he walked in and after exchanging greetings, the Purchasing Officer suddenly said, "Tom, excuse me, I just have to go out. I have to see somebody, but stay her, make yourself at home, there's a box of cigars on the desk, help yourself, I won't be long." So Tom sat there for a while and eventually he thought that he may as well have a look and see what sort of cigars they were and maybe take up his offer to have one. Tom opened the box, and it was completely empty. So he sat there for a while meditating on the fact that he had been offered a cigar and on investigation had found an empty box. Then suddenly the penny dropped. He went through his pockets and found some travellers cheques which he signed and put in the box. He also had some money, some American dollars, which he put in the box and then just sat there waiting for the Purchasing Officer's return. Eventually the Purchasing Officer did return and sat down, very gently opening the lid of the box. He looked inside, and then said, "Tom the contract is yours". That's the sort of wheeling and dealing Tom was very good at.

Lepanto are a fairly substantial mining concern in the Philippines especially in copper and gold as well as being quite diversified in things such as timber, shipbuilding and distilling. Tom had sold a lot of diamond drilling equipment to Lepanto in the Philippines and he somehow managed to talk them into forming a diamond drilling corporation, which was named The Diamond Drilling Corporation, to do not only their own diamond drilling but also outside contract work. This was an excellent idea because not only will your own operation be efficiently carried out but you can get a pretty good idea about what the others are doing which might lead to diversity and joint deals.

On day Tom was in Sydney on holidays and he suggested to me that we might like to set up a factory in the Philippines to make diamond drill bits. There were a few manufacturers there already. We wouldn't be the only ones, but at least Tom would be one of the bigger users and from his point

Fred Velayo, Tom Bain, Tony Taylor and others at the opening of the Manila factory.

of view, I suppose, getting the right type of bits and right type of work was also pretty important. So I spoke with Fred Velayo, the then Managing Director of Lepanto. Fred came to Sydney to discuss things with us, and then later on, I went up to the Philippines to talk things over with them. The deal we struck was very beneficial, as far as Diamond & Boart Australia was concerned, insomuch as we would work with them on a 60/40 basis. Under Filipino law, Filipinos had to have 60 per cent of the equity, whereas outsiders could only have 40 per cent. Provision did exist however for getting a larger equity share after 12 months and if necessary even to buy the Filipino out.

Our part of the deal was to supply the equipment which would come from the Adelaide factory. We had decided to make all drill bits in the future in Sydney in order to concentrate all the work in one area. This had made the equipment in Adelaide redundant, which could now be used in the Philippines. We would also, of course, supply the knowhow and the diamonds required. While we only made a small profit on the diamonds, we also got a royalty and a share of profits. From Diamond & Boart's point of view it was a good deal.

I looked around Diamond & Boart for a suitable candidate to be manager for the Filipino venture, and we found a capable young man in Melbourne by the name of Bob Adams, who had just been married. He had done a very good job in Melbourne selling diamond tools. So we bought him to Sydney and he accepted the job. We trained him in the actual manufacture of diamond drill bits and sent him to Manila where he supervised the building of the factory, which was located right alongside Tom Bain's operation and offices.

We proceeded to make not only diamond drill bits but also diamond saws, as the Philippines produced very good marble. One of the shareholders in this venture was Tony Taylor. Tony had a big marble quarry and he used a large number of diamond saws, not only circular saws, but also gang saws, which are about 12 foot in length. Apart from that they also used quite a lot of what we refer to as thinwall bits. They are steel tubes of various diameters, set with diamonds on the front end, which revolve in a drilling machine producing core, which is then turned on the lathe into various shapes for marble flower stands, lamp shades and the like.

In the beginning Bob had a bit of a problem getting organised with labour, but when we did advertise we always advertised in English, which meant that at least whoever read the ad could read and understand English. Another problem he had was the fact that one of the foremen was the head of a small village, outside of Manila, and he owned a jeep, a four-wheel-drive vehicle, festooned with bells and all sorts of fancy gadgets. This fellow would bring maybe four or five people from his village into work every morning who would have to pay a fee for this service. Well this fellow would go to Bob and say, "Mr Adams, we need another man in this section for the following reasons," and give him a very lengthy explanation as to why this other fellow was needed. He would say, "Well we have just the right bloke in our village, I'll bring him along". So not only would he have another paying passenger in his jeep but he would also demand from that new employee a share of his salary. Although wages are very low in the Philippines, once you get

A rowing 8 presented by the North Shore Rowing Club Sydney, being assembled at the Manila Rowing Club in the Philippines.

your own salary, plus say 20 per cent of five or six others then you are not doing too badly. This is what went on until Bob woke up to it.

The United Nations had a technical school in the Philippines, the head of which was an American and the second in charge an Australian. Bob got to know them both quite well and he would go up to them and say, "Well listen Jack, have you got a good man, a fitter and turner," or whatever. Someone would then be sent along and we would show him a blueprint, a very simple one and say, "Well now how would you go about making this?" If he explained it to our satisfaction, we would then give him a lump of steel and a lathe and say, "Well right, there's the steel there's a lathe now go and turn it". Or, if he was a welder, we would give him a few bits of steel, and say, "Now weld this up for us". When the job was finished, if it was done satisfactorily within the time limit we could be sure that he could do the work quite well. Once he passed the practical test the next thing was to send him to a clinic, because the last thing you want is people who are going to take sickies every so often. So for $4 a certain clinic would turn these people inside out, they would check every part of their anatomy, their blood pressure, hearing, sight, everything about them. Then, if you were satisfied that the man in question was healthy, you'd take him on.

Jobs in the Philippines were desired by many Australians. For instance, Bob had an air-conditioned house with its own swimming pool. He belonged to one of the best social clubs, the yacht club, and the rowing club. He had an air-conditioned car with a driver and work virtually finished on a Friday at noon when all the mining people would gather at the University Club,

Charley Palanco, chairman of Lepanto Consolidated Mining Company opening the factory in Manila.

have quite a few beers and a very pleasant lunch which went on till it was time to go home. Apart from that there were the side trips to Hong Kong, Singapore, Djakarta, etc. From a wife's point of view, in most cases, life could be very boring, the exception being for those wives who had hobbies, such as painting or weaving or whatever or playing bridge or golf, but you can't play bridge or golf all day. At the same time they could make it interesting if they wanted to. The children could be sent to the American school, where most of the white children went. But the standard of schooling is far below that of Australian schools. Once the children return to Australia, they are invariably moved at least one class down.

I attended a few board meetings in Manila, which were rather interesting and held at Lepanto's offices, on the 20th floor of a bank building. They had a magnificent board room and all the directors were paid per diem, that is during the board meetings a young lady would come around and present you with a cheque, but only if you attended.

But the most interesting thing was that about half-way through the meeting food would be put in front of you. Often people tried to answer questions with their mouths full, not quite the correct thing to do. I was rather intrigued by this and asked why it happened. The answer was that it went back to the

Spanish days, when buildings were not air-conditioned. Whenever they had a board meeting, half-way through the afternoon, owing to the oppressive heat, most of the directors would fall asleep. Food would stimulate them and after having a bite to eat they would sit up and take notice again. Although buildings are now air-conditioned the practice still persists.

Later on for some reason Lepanto wanted to reduce their equity in the company and the opportunity was wide open for Diamond & Boart Australia to acquire a much bigger share. The local directors of Diamond & Boart decided that they wanted an outsider, and asked my advice. I suggested Diamant Board of Brussels. I knew they were very large with factories in the United States, Canada, South America and all over Europe, but not in the Far East. So when I suggested that they could be interested the local director and company secretary decided to go over to Brussels. This was a bit of a surprise to me, as neither of them knew anything about diamonds, neither of them spoke Flemish or French and neither of them were business people. Nevertheless an agreement was reached and Diamant Board of Brussels came in on a larger equity than Diamond & Boart Australia, and installed their own manager, a very capable Belgian, who had previously looked after their factory in Athens. Apart from mining bits they also went into oil bits in which Diamant Board specialises as well as all the ancillary equipment that goes with diamond drill bits for the oil industry. I understand that they are doing exceptionally well, have moved to Singapore and have even branched out to Taiwan and are giving the Japanese diamond tool manufacturers a run for their money. I hardly got any credit or thanks for establishing the operation. Not that this particularly worries me, but I am proud of the fact that today it is a very active and successful concern.

Ding Velayo, Fred's brother, owned a bonded warehouse near the airport and when Fred retired from Lepanto he obtained from his brother a nice large air-conditioned office in Ding's sports complex which was near the warehouse. Fred's idea was to take things quietly and look after his own interests and investments. This was fine until Mrs. Marcos walked into his office and said "Fred, you are going to work for me" and this, of course, was an order. Fred, apart from having a degree from Manila University, had a Business Administration and Economics degree from Harvard, he also had formed the largest accountancy practice in the Far East. The salary she offered him was very liberal, even by American standards. So Fred didn't struggle too much. Brother Ding was not quite so lucky. In his bonded warehouse he employed a large number of girls who manufactured hand-embroidered ladies fashion garments under the supervision of an American couple. He also employed a very competent Swiss jewellery designer and manufacturer who would make a ring, brooch or other pieces of jewellery in a soft metal such as copper or silver and then using the "hot wax" process would cast numerous pieces from the original model.

In the Philippines it was the practice, that if you opened a factory of any substantial size and importance you would invite Mrs. Emelda Marcos to come along to open the factory. This would then also bring along jounalists and newspaper photographers to record the event, resulting in substantial publicity. It was also the practice to present her, duly recorded by the press, an article of your manufacture. Ding had a dilemma. Knowing Mrs. Marcos was extremely wealthy and had everything one could imagine he wondered whether to present her with a ring or brooch, maybe earrings — he didn't know. So in his wisdom or otherwise he decorated a beautiful tray with rings, brooches, pendants and all sorts of jewellery and thought, "I'll leave it to her to choose". When the right moment came, he made a short speech and said "Madam, I know it is the practice to present the First Lady of the Philippines with an article manufactured in the new factory". She looked at the tray very intently and said, "Aren't they all beautiful" whereupon she gave the tray to her secretary and said, "Put this into the boot of the car".

I said to Ding, "What did you do?" He said "What could I do, she is the First Lady of the land, I couldn't grab it back from her. When I asked him how much in value was on the tray, he said, "About $170,000".

I was told he went home after the event and cried. The Philippines experience was interesting and educational for one brought up in Australia. it certainly widens one's views on manners, culture and people. Tom Bain and I each presented a paper at the De Beers Diamond Seminar in Manila.

CHAPTER NINE

Successes and Failures

The late 'fifties and early 'sixties saw a tremendous amount of development for Diamond & Boart Products (Australia) Pty. Ltd. One of things we became involved in during that period was the manufacture of diamond saws. These are used principally for sawing very hard substances like marble, granite, glass, ceramics, asbestos and other materials for which conventional saws are unsuitable.

Unfortunately, as far as we could determine, there was only a limited number of people interested and the market showed no signs of improving in the short term. Apart from a few stone-masons, our only other steady client was Wunderlich's, who made the familiar 12 inch tiles used all around Australia in pubs, butcher's shops, kitchens and bathrooms. Wunderlich had a special use for our diamond saws. They found that when they fired their tiles in a furnace to put a glaze on them, they would go out of shape. So in order to return them to a perfect square, the tiles had to be sawed, two sides at a time, using a pair of diamond blades set 12 inches apart and then turned through 90 degrees and sawn again along the other two sides.

Around the time I was pondering the future of diamond saws in Australia, I heard news about an American inventor in California who had developed a special machine for sawing concrete. His name was Cecil Hatcher. Cecil had not received much in the way of a formal education, but had been blessed with the quick understanding and manual dexterity of a natural born engineer. Give him a broken-down tractor, a differential or a gear box, and he would have it working sweetly in no time. When he turned his formidable talent towards solving the problem of sawing through concrete, he came up with a hydraulically propelled machine fitted with a diamond blade that could be raised and lowered by means of a lever and it worked beautifully.

I decided to go to California to see him and was astonished to find that he did not even have a blueprint for the thing. All he had was a drawing on the concrete floor of his workshop which he would consult every now and then as he took up a bit of steel to bend it into shape. I asked for, and was granted, the agency for Australia, and we imported our first machine.

At first, people here were sceptical about the claims we made. "It can't be done," they said. "No machine could be capable of sawing through concrete, particularly reinforced concrete." Fortunately, Shell was prepared to give it a try.

Our first job involved sawing a large concrete slab at a garage at Marrickville to enable a new underground petrol storage tank to be installed. Tom Bain and I went out there and did the job. We sawed through the concrete which they were able to lift out in one piece, giving them access to the old tanks. When the installation had been completed they simply hoisted the slab back into its original position.

Clearly, the potential uses for the machine were enormous, especially for local government councils and the like. With so many services like water, gas, electricity and telephones being located underground, beneath public footpaths and roadways, any necessary repair work always meant smashing up the pavement with jackhammers, creating havoc for pedestrians and motorists alike. So I went along to Marrickville Council trying to drum up some interest in the new machine. As it is with a lot of new ideas, they were unconvinced until we had given them a free demonstration.

In 1959, we formed a company called Concut Concrete Sawing and Drilling Pty. Limited. Tom Bain was its first manager and a German diamond cutter, called Carl Emmel, was its first employee. Concut took of so fast, and with so much success, that after a very short time most councils around Sydney prohibited the use of jackhammers on footpaths and insisted on them being diamond sawn.

While the sawing of old concrete produced a tremendous saving in time and effort, another equally significant development was the sawing of green concrete within about 12 to 36 hours of its being laid.

A large slab of concrete made from a mix of cement, sand, aggregate and water and measuring more than say 12 feet square, will when it contracts, develop ugly cracks as soon a water evaporates out of it. This can be overcome, however, by pouring the concrete continuously, as they do in the United States, and sawing a contraction joint through its surface down to about a quarter of its overall depth. By doing this, a weakened plane is produced allowing a contraction joint to form beneath this sawn joint made by the diamond saw. The technique allows long stretches of unbroken reinforcing steel to be used, greatly increasing the strength of the finished slab, and also prevents the sort of movement of slabs common in concrete roads. With a lot of old concrete roads around Sydney, each slab moves slightly as you drive over it. When water gets down in between the slabs, the surface pressure of the traffic produces a kind of pumping action which, in time, causes the surface to become uneven. You used to be able to notice this effect quite clearly when travelling along the old concrete road to Newcastle. The slabs were all at slightly different levels so that when the wheels passed over each joint in the concrete, a distinctive clickety-clack was produced.

The management of Concut in California changed, and I found I could not deal with the people who took over. Even if we had been able to reach some agreement enabling us to manufacture under licence here in Australia, the kind of royalties they were talking about would have been prohibitive. Eventually we parted company and started to design and build our own machines for the Australian market. They have been built ever since right here in Australia and adapted to a wide range of uses. We built one very large model that carried a 56 inch blade for use on the concrete taxi ways of Darwin Airport where the slabs were up to 16 inches in thickness.

Carl Emmel, Concut's first employee, sawing Essendon Airport.

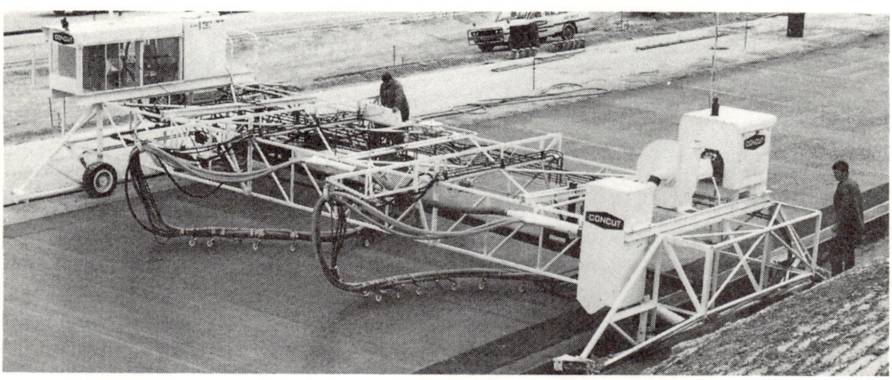

Sawing contraction joints

Then came the thinwall diamond drill bit used for drilling holes through concrete. This innovation made it possible to accurately drill holes of the correct diameter and in the precise spot required. These holes were made for the installation of gas mains, conduit pipes and so on. Nowadays, thinwall bits are made up to 24 inches in diameter and larger.

The technique has been used to good effect in the United States where, in the construction of apartment buildings for instance, they leave the service holes to the very last. This allows each new purchaser some flexibility in choosing where they want the kitchen stove or bath located. The holes can be easily drilled through the walls to accommodate the various service pipes. It is a good selling point when prospective purchasers are being shown through.

Probably one of our biggest jobs in those days was on the Chevron Hilton Hotel at Kings Cross. During construction, either the builder or the architect forgot to allow holes for the fire sprinklers, and we got the job of drilling something like 2,200 holes through the various floors of the hotel to accommodate them.

Perhaps the most complex job was one we did for the Department of Mines in South Australia. They wanted to drill five holes next to a wharf so that special steel girders known as "dolphins" could be put in place. The dolphins served to steady ships which were docked at the wharf, allowing trucks to be driven on and off safely. The problem was that the drilling had to be done through some of the hardest granite I have ever seen and in 30 feet of water no less. The Department of Mines discussed the problem with us and sought our advice. They wanted the holes to be 18 inches in diameter and go down to a depth of 9 feet. In the end, they decided to try it themselves and had the special bits made up by one of our competitors. I thought it was rather rude of them at the time, but they came back to us in the end when they ran into trouble. The bits they were using couldn't do the job. I think the basic problem was that the tubing used had been welded and developed a definite banana-shape. So we made up new bits and devised a method for carrying out the actual drilling.

At one point I went down with a set of aqualungs to check on progress. I probably made my descent too rapidly because, still recovering from the effect of a head cold, I finished leaving one of my eardrums at the bottom of the harbour. Diving in those water could be hazardous in any event. That area of Australia is notorious for sharks, and the presence of tuna fishermen in the area, cleaning their catch and dumping the offal into the water helped to keep us pretty alert. Technically, it proved to be quite a challenging project and there was a fair bit in interest at the time in mining circles. A Mines Department representative in fact later read a paper on the project at the Broken Hill Drilling Convention.

Another interesting job we were asked to do involved drilling a number of 6 inch holes through bank vaults. Not long before, one of the suburban branches of a Melbourne bank had been held up by a couple of bandits. After helping themselves to the money, they ordered the staff into the vault, locked the door, and walked out. When the bank was found deserted, somebody fortunately decided something was wrong and rang head office. It took a while to get someone out to open the vault but finally the employees

Tom Bain at Port Lincoln with the 450mm diameter diamond drill bit, and 3 metre core barrel used for drilling in 30 feet of water at Port Lincoln harbour.

were released safely. Another two hours, however, and they would have all been dead through lack of oxygen.

The incident sparked off a rush of orders from banks all over the country wanting ventilation holes put in all their vaults. It was no mean feat because bank vaults are invariably encased in very thick reinforced concrete. It meant drilling a 6 inch diameter hole right through the wall of the vault and installing a hardened steel plate over it on the outside. The plate could be moved only from within the vault by means of a bolt. The idea was that anyone locked in the vault could undo the bolt and let air in through the ventilation shaft. More air could be sucked in by means of a vacuum unit within the vault if this was required. I don't know how many banks we ended up doing, but it must have run into many hundreds.

Another time we were asked to drill a hole in the bottom of the harbour near Garden Island. Tom Bain and I wanted to make a careful inspection of the area first so we each donned aqualungs and went down to have a look. To reach the intended site we had to swim beneath a barge, the underside of which had a thick encrustation of barnacles. Foolishly I got too close and scraped my back quite badly on its razor-sharp surface. The water seemed to be full of blood. It made me very nervous indeed, knowing the attraction blood has for sharks. Tom and I certainly didn't linger too long over that inspection.

We eventually drilled the hole but it proved to be quite a difficult job. The

The core at Port Lincoln.

harbour bed at that point was covered by a very thick layer of mud. Once we disturbed this with our drilling operations the water became so dirty it was impossible to see more than a yard or so in any direction. But at least we proved that we could do it.

Next came a very important development which allowed a more accurate means of testing the foundations for buildings and other structures. On big construction projects, testing the concrete foundations is of critical importance. The architect will usually call for samples of the concrete that has been poured to determine how it will stand up to the enormous stresses that great buildings and other structures impose. The new capacity to drill clean and accurate holes in concrete allowed samples or "cores", usually of 6 inches in diameter, to be taken from various sections in the foundations during construction. These could then be subjected to various pressures and other tests under controlled conditions to determine their suitability for the purpose in question. It was amazing how quickly architects and engineers in Australia tumbled to the advantages of this excellent method.

Later on we developed a special machine incorporating multiple blades capable of cutting concrete surfaces. The machine was designed to cut grooves in airport runways often using up to 64 blades at one time for this purpose.

From studies made of aircraft landings, engineers had noted a certain aqua-planing effect occurred on wet runways. Cameras strapped to the undercarriage of various aircraft, showed that the tyres would spin at quite high speeds on touchdown, but slow down dramatically and eventually stop even while the plane was still moving at comparatively high speeds.

This aqua-planing effect is caused by an accumulation of water forming between the tyre and the concrete or bitumen surface of the runway, and can occur even at speeds as low as 30 miles per hour. It can be very dangerous because the pilot virtually loses all control. I know of one incident where an aircraft went off the runway in Hong Kong and ended up in the bay. There was another case in America where a plane just kept going off the end of the runway and went through a building destroying both the plane and the building.

In Los Angeles there is a complex, spaghetti-like network of highways linking the city with outer suburbs. As a safety precaution, miles and miles of the freeway system has been grooved. There, an accident at speeds of 70 miles per hour or so can involve many cars. The cost, not only in terms of human life and suffering, but in property damage, lost wages and productivity, can be enormous. The Americans have adopted a very pragmatic approach to the problem opting to spend quite large sums on preventative measures, rather than casting the greater burden on insurers, and ultimately on society itself.

Road grooving generally involves cutting grooves one to one and a quarter inches apart and about a quarter of an inch deep. Concut built their own machines for this purpose, which proved highly successful, as well as much of the ancillary equipment needed in this type of work.

One of the big jobs we took on was Kai Tak International Airport in Hong Kong. Hong Kong Airport runs out into the bay, and after heavy rains it can be quite a problem for a pilot to pull up his aeroplane at the end of the runway. Concut put grooves in at right angles to the runway itself, so that the tyres would squeeze out to the sides any water lying in the grooves. For

Grooving concrete runways

that particular job, the saws used had tapered or wedge-shaped diamond segments.

Later on, I introduced concrete sawing and drilling in New Zealand where we were asked to drill a number of holes through very thick concrete on the apron area of Wellington Airport.

The next major development for us was in grinding concrete and we had a big contract involving work at Tullamarine International Airport just out of Melbourne. The specifications for the tarmac area required the surface to be laid flat to within a quarter of an inch. I remember thinking at the time when I read them that they would have a hard time finding someone willing to take on a job of that sort which demanded such exact levels. A company did take it on, however, an Australian/American company, but of course, they found it virtually impossible to keep within the specified tolerances. The Department of Works came along with a 12 foot aluminium straight edge which they put down, marking the surface here and there with various coloured paints where high spots had to be ground down. That's where we came in. We used a flat blade with diamond impregnated segments attached, very similar to a lawn mower, and ran over the high spots grinding them down. It worked very effectively.

Next came the drilling for the runway lights. The contract called for holes to be drilled into the concrete with grooves cut running between each light so that the conduit could be laid.

Apart from all these applications, Concut has also been called upon to do a lot of wall sawing, particularly in areas where openings have to be cut into a wall for doors or windows or whatever. The technique is also used in cases where a foundation might have sagged, producing ugly cracks in the wall.

Concut developed rapidly under the management first of Tom Bain, then David Rees and more recently David Pirie. By the time I handed it over to them, we were in every State in Australia, and, before I left, I had established a branch in Hong Kong in conjunction with Pioneer Concrete. The Hong Kong operation involved quite a bit of drilling work connected with the construction of stations for the underground railway network. The last time that I was in Hong Kong, in fact, a new underground station, the size of a six storey building, was being constructed in Kowloon. Concut today operates in every large city in Australia, and all their vehicles are radio controlled. A lot of diamond drilling machines which they use today, particularly for heavy duty applications, are hydraulically operated and designed and built by Diamond & Boart. The company has certainly come a long way since its formation in 1959. I am proud to have been associated with its early years especially since it was responsible for the introduction of concrete sawing and drilling techniques in Australia and New Zealand.

Around 1960, Pilkington, one of the largest plate glass manufacturers, asked us to make a machine that would automatically grind the windows of motor cars, which was a large part of their business. At the time such machines were made only by one company, in the United States. So we designed our own Australian machine for this particular purpose. However, Pilkington's head office in England insisted that they buy American machines, which had been and were being used all over the world. It was only through the influence of their Australian manager, Alf Burrell, that

Automatic pencil edging machine for the automatic grinding or pencil edging of car windows at 350 per hour, designed by Diamond & Boart Pty. Limited.

Pilkington's in Victoria gave our Australian machine a go, which proved to be a great success.

The Automatic Pencil Edging Machine, as we called it, was a large machine, measuring nine feet by about five feet, and weighing about seven tons. Its function was to grind the glasses for motor car windows automatically. As it turned out it was able to produce 350 glasses an hour whilst the American machine's output was only 180 an hour. The machine has been in operation in the Villawood plant since 1961, and is still working, sixteen hours a day, seven days a week.

Later, two more such machines were built by us with an improved design, one for the Geelong factory and one for the Kilkenny factory in Adelaide. We also received orders for an automatic machine to be sent to Poland, and for four machines to be sent to Russia, whilst another two went to England and another to South Africa.

The game of bowls has been popular since Sir Walter Raleigh's days

They used to be made from ligum vitae, which is a very hard and heavy wood. In fact it sinks in water. Block and tackle on the old sailing ships were made from it and later on it was used as bearings for shafts. Ray Hensell hit upon the idea of moulding the bowls using bakelite. Bakelite powder however is very abrasive and a tungsten carbide lathe tool, which is one of the super hard powder metal tools was used to turn the bowl after moulding. After taking a roughing cut the tungsten carbide tool would have to be taken out of the lathe and resharpened on a diamond impregnated wheel, before the finishing cut could be taken. When this was pointed out to me and after visiting the factory and seeing the application I realised this was the ideal application for a diamond lathe tool. So I make a lathe tool with a negative rake and radius from an Australian Copeton diamond. The tool did 960 bowls before it came back for resharpening. The follow-up was that I made a round diamond tool clamped in a tool holder which allowed Hensell to turn a number of bowls and when that side was blunt all he had to do was undo a nut, turn the diamond a little, retighten the nut and carry on until the whole circle of the tool had been used. This then produced 3500 bowls on the average. On each end of a bowl there are a number of grooves so for this application I made a number of identical radiused diamond tools which were plunged simultaneously into the bowl.

Both ends of a bowl displays on ensignia and this is engraved into the bowl using a diamond engraving tool mounted to a pantograph machine. One can imagine what a difference all this made to Hensell's production and accuracy. It was from then onwards that he experimented rapidly and became a large exporter to numerous countries all over the world.

A large manufacturer of sporting goods, approached me advising me they were going to manufacture bowls and compete with Hensell. They wanted us to manufacture the required diamond tools, however due to our lengthy association with the Hensells there was no way we would make tools for them and I told them so. Their project was dropped.

It was an all-Australian effort and credit is due to the late Ray Hensell who was later on ably assisted by his sons Bruce and Graham who have expanded the business greatly. Today polycristalline diamond tools are used, which are millions of diamond particles fused together.

In 1955, when I happened to be in Chicago at the Diamond Tool Convention, General Electric of America announced that they had been successful in manufacturing synthetic diamonds. It was indeed an incredible achievement. Up till then scientists had known that only terrific heat and pressure could crystallise carbon into diamonds. Heat can be produced and so can great pressures. But, in order to be able to combine both terrific heat and pressure, it is necessary to design and manufacture a metal that could withstand the necessary pressures under terrific heat.

In 1776 a French scientist, Lavoissier, discovered that the element carbon crystallises in two main allotropic forms, namely diamond and graphite. The fundamental difference lies in the arrangement of the carbon atoms in the crystal lattice as a consequence of which both modifications show great differences in a variety of physical properties. For example a diamond is usually a transparent crystal and is the hardest substance known to man. Graphite, on the other hand, is completely opaque and exceptionally smooth and soft. Since Lavoissier's momentous discovery, numerous scientists had attempted to make synthetic diamonds. For example Hannay in the nineteenth century experimented for a long time with carbon or graphite as a base using terrific heats and pressures. He never succeeded and although there are two diamonds in the British Museum supposedly created by Hannay, it is rumoured that two of his young helpers bought two rough diamonds at Hatton Gardens in London, which were introduced into Hannay's experiments. Since then it has been proved by X-ray, that the two diamonds are of different origins having slightly different atomic structures.

Then came the French scientist, Moisson, who in 1893 tried to emulate meteorites, which are known to sometimes contain diamonds basically of the same substances as the earth, namely, iron, carbon and nickel. In fact some of the diamonds found in northern Canada, near Hudson Bay, were deposited by huge meteorites which crashed to earth there thousands possibly millions of years ago. It was assumed that, when a meteorite is attracted towards the earth by the earth's magnetism, the outer layer solidifies first forming a steel band. The inside of the meteor, still in the form of molten lava, creates gases which together with terrific heat and pressures crystallise the carbon into diamonds. So Moisson obtained a quantity of iron into which he introduced carbon. He then heated this up and plunged it down a well. He claimed that very minute diamonds formed on the outside, however this has never been substantiated. Moisson then hit upon the idea of plunging the heated iron and carbon into molten lead, which meant that it would cool more slowly and thus produce larger crystals. This he did, and he claimed that larger crystals were formed. Again, however, this has not been proved. Between 1914 and 1918 Sir Charles Parsons experimented with Hannay's methods, and later wih Moisson's methods, finally introducing a few of his own. Parsons claimed, after spending a considerable fortune, that neither Hannay or Moisson nor he had ever produced synthetic diamonds.

So you can imagine the great breakthrough that occurred in 1955, when General Electric announced that they were able to make synthetic diamonds. General Electric is still producing them today, and those synthetic diamonds have taken their place in the diamond industry. Synthetic diamonds are used as an abrasive in diamond impregnated wheels, used primarily for grinding tungsten carbide, ceramics and glass. The first people to produce synthetic diamonds were A.S.E.A. of Sweden, a large electrical and motor generator manufacturer. Later De Beers started manufacturing and experimenting at the Diamond Research Laboratory in Johannesburg and now have a large factory in Shannon, Ireland. I understand that Norton of the United States, and Philips of Holland are also able to manufacture synthetic diamonds. As well as a firm in Japan, the Russians and even the Chinese, have made synthetic diamonds. So there are quite a number of people who are able to take carbon, introduce a catalyst, such as iron, or cobalt, nickel and/or other minerals, subject this to terrific heat and pressure, and produce very small diamond crystals.

One day Professor Ted Ringwood of the Australian National University contacted me and said that he had made synthetic diamonds to win a bet. Apparently he was at a cocktail party one Saturday and somebody bet him that he couldn't make a synthetic diamond. ''Well,'' he said, ''how much time do you give me?'' They said, ''Well, till next Saturday''. He had been able to produce synthetic diamonds, which I have seen, in a week.

We toyed with the idea of making synthetic diamonds in Australia. We went into various costings and so forth, but found the market in Australia to be too small. Also the competition from the De Beers Company and General Electric would be tremendously fierce. There is no doubt, however, that synthetic diamonds could be made in Australia, particularly in view of the fact that we have a tungsten carbide manufacturer here in this country, Titan, a subsidiary of B.H.P. at Newcastle, is capable of producing tungsten carbide. The only other requirement would be a huge press, of about 10,000 ton, which are made in Sweden and the United States.

During my visit to Chicago in 1955, I met a diamond tool manufacturer by the name of Lep Metzger. Lep had been a gramophone record salesman, who used to sell records door to door. Then he met a central European person who manufactured diamond impregnated wheels. Though he made very,

very good wheels he was unable to sell them as he was not a salesman. So Lep and this man got together and formed a company called Supercut. Supercut was very much advanced in the manufacture of diamond impregnated wheels, particularly for the glass trade and made most of the diamond wheels for the Sun Tool Company's pencil edging machines. Supercut also made a great number of diamond saws, for which there was a very large market in America. The factory occupied a whole block.

Lep told me the story of how he was robbed one day. He had a foreman who displeased him so much that he sacked him. Lep was always extremely conscious about security. He had a burglar-proof alarm system and a nightwatchman, who used to come around every hour, pick up the phone, quote his number, take a quick look around, say "All's well", and then ring off and go on his way. So when Lep sacked the foreman, he took the trouble of changing the combination to his safe, but like a lot of people his memory was not infallible so he wrote the combination down on a bit of paper. One morning he came in, picked up the phone, quoted his number, walked to the front door, opened the safe, and there was nothing in it. It had been completely and utterly cleaned out. He rang the police and the insurance company and advised them of it. Before long the place was swarming with detectives and insurance people wanting to find out what had happened.

Well, what had happened was this. The foreman still had the key to the back door and although Lep had changed the lock on the front door, he forgot to do so on the back door. Knowing the set-up, the foreman came down one night, heard the nightwatchman come in and brake the seal. He quietly opened the door and snuck into the shadows waiting there until the nightwatchman had rung off, said, "All's well", and closed the door again. The foreman then had one hour to do whatever he wanted. He went to the secretary's desk, opened all the drawers, and right at the back of one underneath some documents found a tiny slip of paper with various numbers on it. He realised that this was the new combination to the safe. Wearing gloves, he turned those numbers, until the door opened. He then took every diamond that was in the safe, put it in a bag, carefully locked the safe again, scrambling the combination. He then waited for the nightwatchman to come along and break the seal. As soon as the nightwatchman had broken the seal he snuck out via the back door, carefully locking it and went on his way. The perfect crime.

Once the insurance company discovered that Lep had left the combination lying around they pleaded negligence on his part and wiped their hands of the matter leaving poor old Lep to carry the cost. The only fingerprints on the safe were his, as the former foreman had worn gloves. Lep was desperate. He didn't have a diamond left, he had orders for tools, and a lot of people were whispering amongst themselves that he'd done it himself hoping to claim on the insurance company. The only people who had confidence in him was a firm in New York, Bernard Jolis, that he used to deal with which sent him more industrial diamonds and so he was able to start again.

But poor old Lep went grey, until one day, about 18 months later a man in the industrial diamond trade, rang Lep, and said, "Lep there is a fellow down here, he's showing me industrial diamonds, he obviously doesn't know anything about them, they could be yours." Lep said, "Well keep him there, keep talking to him. Pretend you are examining diamond''. So Lep rushed down, grabbed a couple of detectives from the police station and raced up to this man's office. Sure enough they were his diamonds, and the man was apprehended.

What had happened was that the foreman whom Lep had sacked, and who had stolen the diamonds, went home straight after. In a lot of the homes in America, as in Europe, they have cellars. The foreman had gone down into the cellar of his house with his wife, prized a few bricks out of the wall, put the diamonds there and restored the bricks. He then said to his wife, ''We will leave them down there until the whole thing blows over in a year or two's time. In the meantime I'll get myself another job and say in two or three years time we'll take the diamonds to Canada, sell them there and we'll be on clover.'' But there was one thing he did not reckon on and that was that his wife should have a boy friend who would get her into trouble. In those days (in the United States) an abortion cost a lot of money and the boy friend didn't have that sort of cash. So the wife said, ''No problem, come down to the cellar with me, we'll prize the bricks and get the diamonds out. You go down and sell them and when you get the money, I'll go and see a doctor.'' So the moral of the story is, if you're going to steal diamonds, make sure your wife hasn't got a boy friend who is going to get her into trouble. Either that or hide them in a place she doesn't know about.

Although Diamond & Boart was booming in 1969 and going ahead in leaps and bounds, (in fact our accountant referred to me as an Empire Builder) the problems that I had had with Longyear and Christensen had worried me to such an extent that I considered selling out. So I approached Diamont Board, Brussels a very large industrial diamond tool company owned by the Societe Generale, (a large Belgian conglomerate) which is related also to the Central Selling Organisation. I think the big mistake that we both made was that I should have gone to Brussels and talked it over with them, or they should have come to Australia and talked it over with me here in Sydney. Letter writing is never satisfactory and never leads to anything. You have to face people and talk to them face to face.

My father had passed away in 1963, which was a tremendous blow to me, and I was left to carry on on my own. I couldn't go to him for help, advice, or support so the next best thing was to go and see George Lane. He was a very well-known solicitor to whom my father used to go to for advice on any legal problem that he had. So, I went to see George, who was also chairman of O.P.S.M. Industries Limited, and Paynter & Dixon Industries Limited, the big industrial builders. I explained the situation to George in detail and he listened very intently. Then he asked for two days to think it over. When I went back he said, ''Well, I am as you know, chairman of O.P.S.M. and Paynter & Dixon. How would it be if these two companies, which are all-Australian owned, took over on the same terms and conditions as Diamant Board of Brussels?'' So I said, ''That's fine, that's marvellous, I'd love the company to be in the hands of two Australian companies, if I'm going to quit, in spite of the fact that my parents come from Belgium.'' So the company was sold to both O.P.S.M. and Paynter & Dixon with the latter company taking two thirds and O.P.S.M. one third. Both of these companies used diamond tools. O.P.S.M. for grinding their lenses, and Paynter & Dixon through Concut, (a fully owned subsidiary) for sawing and drilling holes through concrete.

I was asked to stay on as managing director for three years which turned into eight. But from that moment on, a change came over Diamond & Boart and Concut. In the past when my father was still alive I hadn't noticed that whenever I went overseas, certain people at Diamond & Boart would get together and whisper in my father's ear about this person or that person being no good, and that they alone were the best and had the company's interests at heart. I did remember receiving a letter from my father when I was in Antwerp, which denounced certain people, who were in fact doing a tremendous job in the company, because Dad didn't know better. This really retarded the progress of the company enormously. I knew that when I got back from overseas I would have to try and straighten this all out. There were still others who were at or near the head of the company who didn't want the company to grow any bigger because they feared greater responsibility or realised their shortcomings.

Things also changed in other areas. When I was in charge, and owned the whole of Diamond & Boart, there were never any board meetings. I just used to call various people into my office and discuss things with them. I'd just talk things over with people in accounts, manufacturing or marketing. We would just sit round the table and talk things over, formalities had no real place. The company was extremely well established and doing remarkably well. Apart from having an office in Mount Isa, we had an office in Brisbane, a factory in Sydney, an office and building in Melbourne, a factory in Adelaide, an office in Perth and in Kalgoorlie and later on, an office in Singapore and Kuala Lumpur. This system had worked well.

The moment the new owners took over, we had to have monthly board meetings, where everybody sat round the table and discussed various matters. I suppose it is difficult when someone owns a company, has built it up and has made all the decisions, whether right or wrong, to work with a board. The other thing is that with a board consisting to a large extent of young men, you will often find one or more ambitious young fellows who will try to jockey for the top position by easing out the older man when the latter doesn't own the company anymore. My mistake was, perhaps, to stay on as general manager. I should have left the company after the initial three-year period which was agreed upon when I sold out instead of staying on for eight years.

Even staying on as a director can have its problems in so much that, having owned the business before, one can be over-zealous or worse still, possessive, which can lead to, or one can be accused of, interference in management.

Though the late 'fifties and early 'sixties were years of great expansion and development, for us they were also sad years. My mother passed away in 1960, a great blow to my father, who died three years later from a heart attack. But, he left having passed on his great knowledge of precious stones, and the diamond business, to the sons and daughters of this young country he had come to know and love so well.

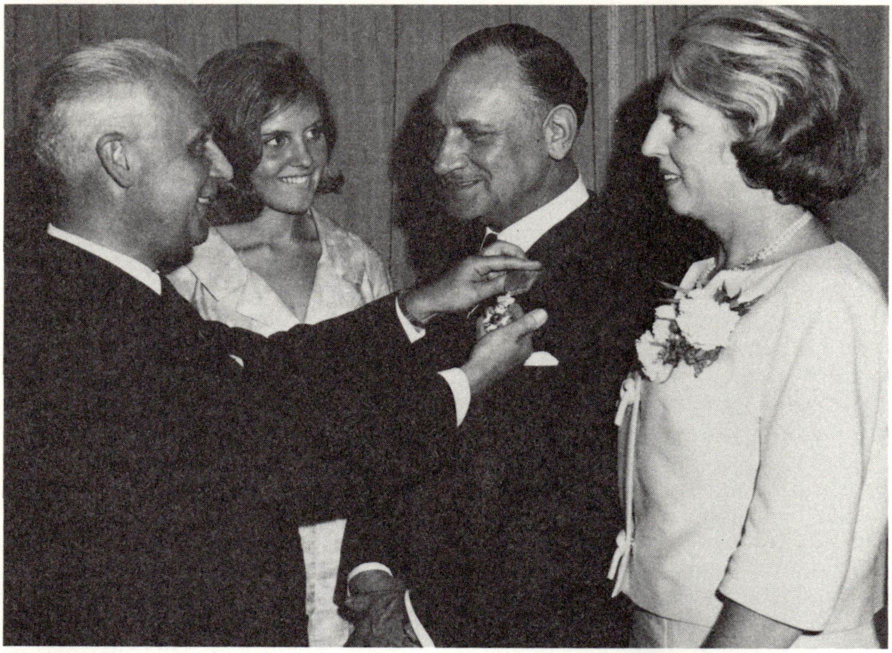

His Excellency the Ambassador of Belgium, Willy Van Cauwenbergh, Vivienne Joris, Albert Joris and his wife Margaret Joris, on the occasion of Albert Joris being made a Knight of the Order of Leopold.

CHAPTER TEN

A Conflict of Interests: The Stellar Story

Bill Walkley, later to become Sir William Walkley, started off as an accountant and company secretary of a small petrol company in New Zealand. For a man with big ideas, like Sir William, New Zealand was too small. So he decided to come to the big smoke, he came to Sydney to start a petrol company here in Australia, an all-Australian company. Obviously the man had tremendous guts because this was in the Depression. I believe he even went around knocking on people's doors trying to sell a few shares in this newly formed company, Ampol.

Well, can you imagine what the large companies thought about all this, people like Shell, Plume, Vacuum, Mobil, and others. To start in opposition to such people during the Depression certainly took a tremendous lot of effort and a tremendous lot of guts and as Mr. Fraser would say "Life is not meant to be easy". But he certainly made a success of it. Later on, as the company grew, he also became a great promoter of sports and many leading sportsmen were brought to Australia as a result of his efforts. When we still had petrol rationing, just after the war, he managed to break petrol rationing by bringing in a couple of tankers loaded with petrol from France.

Sir William made 27 trips to the United States trying to find sufficient money to drill for oil and gas here in Australia. Out of those 27 times, 26 times the door was closed in his face and he was told not to waste their time. But on the last trip, the 27th, he found somebody in California who had some money and was prepared to take a gamble. Sir William had always believed that there was oil and gas in Australia and so Ampol Exploration was formed.

I remember my wife and I travelling with Sir William and Lady Walkley on a ship from Europe. At the time he asked me, "Albert, do you want to buy a few of those Ampol Exploration shares?" They were about five shillings at the time and they had started to drill about 120 miles from Perth. I remember saying at the time, "I'll wait until you have drilled the first hole, which will be a dry one, the shares will then drop to about three shillings and I'll buy them." Famous last words. They struck oil in the first hole drilled and the shares shot up to seventeen pounds ten shillings. If I'd heeded his advice I could have made a fortune, but I suppose that's how the cookie crumbles. But it proved one thing, that we did have oil and gas in Australia and although the first hole drilled was not economical, at least it proved the point and the story from then on at Barrow Island is history.

Very few people realise that Sir William was not only responsible for starting the oil and gas industry in Australia but was also responsible for the search for diamonds and eventual discovery of diamonds in large quantities in Western Australia. This is how it happened. After Sir William retired from Ampol he remained a very active man and couldn't sit in the one spot for more than 5 seconds. I recall that he phoned me at home, about 4 or 5 times one Saturday, and said, "Albert, are there diamonds in Australia?" I answered, "Yes". He then asked, "Where are they found?" I answered, "In every State in the Commonwealth". He asked, "Where have most of them been found?" I answered, "In New South Wales". "If you were to go and look for diamonds Australia-wide, where would you go looking for them?" he asked. I answered, "Western Australia". "Why Western Australia?" he asked. "For the very simple reason that geologically Western Australia has not been disturbed, like New South Wales. Furthermore, if you believe in the Dr. Wagner theory of drifting continents, well then Western Australia was that part of Australia which was joined to Africa."

So, the next thing I knew we were on a plane to Perth. When we got there we walked into the Ampol office, he grabbed hold of a geologist by the name of Murray Johnson and asked him, "Murray where do you find diamonds in Western Australia?" and the poor man scratched his head, looked up at the tectonic map hanging from the wall, and said, "Well I think I would start up here Sir William", pointing to an area south of the Fitzroy River, which later on became well-known as the Noonkanbah Station area. And the next question, "Why up there?" "Because there are lamporite pipes there." "What are they?" He answered, "Well, they are related to kimberlite pipes".

The next instant we were in a taxi on the way to see Professor Prider in the Geology Section of the University of Perth. Sir William asked, "Do you have a geologist who knows anything about diamonds?" To which Professor Prider replied, "We have a Doctor Jeppe here from South Africa who has had diamond experience there". Then Professor Prider said, "He has a twelve months sabbatical leave coming up". Next thing we were in another taxi to see Doctor Jeppe. We asked whether he would work for us. He said that he would, at least during his twelve months sabbatical leave. So Stellar Minerals Pty. Limited was formed in May, 1968.

The company consisted of Sir William Walkley, Alexander Pongrass, Lady Walkley, George Lane and myself. George Lane was later replaced by his son David. We then organized equipment for Doctor Jeppe, four-wheel drive vehicles, tents and camping equipment, and off he went.

Alluvial diamonds had been found in the Nullagine area, which is in the north-west of Western Australia. In fact Doctor Jeppe had shown me a few he had which came from that area. They were quite nice looking diamonds.

There is also a very interesting story which I have every reason to believe is true. It seems that in 1896 an Englishman, from a very well-to-do family, David Carnegie, led an expedition across the Simpson Desert. One of the members was George Stansmore. Whilst the party was within a couple of days of Halls Creek, towards the end of the expedition, Stansmore was

The Stellar mining camp at Noonkanbah in Western Australia.

cleaning his rifle and accidently shot himself. The others searched his body for papers, identification and other personal possessions, which they sent to his family in England. Carnegie wrote a book about the expedition and made reference to finding something of great interest in the desert.

It was in the late 1960s that Norman Stansmore, a great nephew of George Stansmore, a member of the expedition told a friend of mine Rees Towie. (Who later became chairman of Northern Mining Corporation) that that something of great interest found in the desert was a diamond, which I believe weighed about 25 carats.

As prospecting for diamonds is an expensive business, especially in the outback of Australia, where everything has to be carried by plane, or over a long distance by four-wheel drive vehicles, I suggested to the board, that in order to earn a cash flow, we ought to mine sapphires in New South Wales. I know the Inverell and Glen Innes area very well and was familiar with sapphires. Also being in the diamond business and having done some diamond mining myself I was also familiar with the mining techniques involved. To support a team in Western Australia with its tents and food, drilling machines and front-end loaders, trucks and so forth, plus taking up leases for sapphires here in New South Wales, ran into a lot of money. So it was decided to form a public company. That is how Stellar Mining N.L. became the first Australian diamond explorer to be floated with a total of 1,250,000 fifty cent shares which were issued at par, together with 3,137,500 contributing shares paid to 15 cents, with 1 cent payable on application and 14 cents at call. Stellar Mining N.L. had a nominal capital of $10,000,000 divided into 20,000,000 fifty cent shares. The directors were Sir William Walkley as chairman, Alexander Pongrass of Pongrass Industries, Mr. Glen

Drilling Lamporite pipes in Western Australia.

Shields a very highly regarded and qualified mining engineer, Mr. David Lane, a solicitor, and myself.

I was doing a bit of flying in those days, and I took a plane from Bankstown and flew it up to Glen Innes, and then over an area about 12 miles north of Glen Innes over Reddistone Creek. Some of the best sapphires in Australia come from this particular area. I landed at Glen Innes and motored up to Bullock Mountain to the area I had flown over and there negotiated with a German prospector who had a small sapphire area. I was particularly interested in the fact that Reddistone Creek in this particular area widened into a very nice billabong with two granite bars running across it at an angle of about 45 degrees. In any case I didn't mention anything to the German about this and negotiated with him, eventually buying it for $7,000.

Then later on I did the same at Inverell. I flew over an area, landed at the aerodrome about 5 kilometres out of Inverell, and went by jeep over an area at Horse Gully, which belonged to two farmers, Newmarsh and Moses. The asking price was $200,000 for 35 acres, and I came back to Sydney and suggested to the board that for that sort of money we could buy a whole farm of about 2,500 acres, a farm with a very nice house on it and graze a few head of cattle on that area where no mining was done. We could then mine the area for sapphires, and use the outbuildings to store machines and equipment, and the house for sorting and grading the sapphires.

I suppose we all make mistakes. The chairman would not entertain the idea of the farm, and said, "We're not bloody farmers". I still recommended the area and subsequently the company bought only 35 acres for $200,000, which when completely worked out had to be returned to Newmarsh and Moses. When I saw Newmarsh and Moses some time later, I asked them if they would have sold the whole property for $200,000 and they answered, "We wouldn't have let you out of our sight, we would have jumped at it." A pity, but these things happen.

As mentioned earlier, Jeppe set-up his camp at Noonkanbah Station, and we were drilling extensively. Not once did we see an Aborigine. Later on, when a large American company wanted to drill in the same area, there came a great cry of "sacred ground" and Aborigines came from far and wide with Mr. Hawke's son in tow, at one stage. Anyhow, Jeppe went up further north and crossed the Fitzroy River, taking up areas at Ellendale which are now producing the most beautiful diamonds that I have ever seen. In New South Wales, Stellar took up various areas at Oberon, which has produced some very beautifully coloured though small sapphires, as well as an area at Delegate near the Victorian border. In the past we at Jules Joris Pty. Limited had bought beautiful diamonds which came from Beechworth which were found there in the old gold rush days. They must have originated somewhere, and as Delegate had produced a lot of diamond indicator material, we took up an area there.

In September 1969 I made a statement, which was reported in the newspapers, about Western Australia, where I said, "If diamonds were found in the outback of Western Australia, there would be no great transport problem, as there is with base metals," and I went on to say, "If a big diamond field was found in Western Australia, it would have a dramatic effect on the nation's economy". We shall touch on that later.

In the meantime, we started mining sapphires at Horse Gully, which was producing very nicely and I took up premises for Stellar at Brookvale, not very far from Diamond & Boart. We had a number of girls there to do the

Arrawatta sapphire mine at Horse Gully, Inverell N.S.W. next to Stellar mine.

sorting and grading. The heavy media separator, which I ordered from Mitchell Cotts in Western Australia, was operating exceptionally well on sapphires and I think it was one of the first heavy media separators that was used in Australia. The concentrate from it came down to Sydney in sealed drums and was sorted at Brookvale.

In the meantime, Sir William Walkley had retired as chairman, and I was appointed chairman of Stellar even though I had previously been appointed chairman of Audimco, a company which had a licence to search for diamonds, tin and zircon. Then, suddenly, about twelve months later, Alex Pongrass started saying, "Conflict of Interest."

An extraordinary general meeting was held, which we both attended. It created quite a lot of interest at the time. What staggered and disappointed me was the fact that Sir William and Lady Walkley had sold their shares and rights to both Alex and George Pongrass. This meant that, although at the extraordinary general meeting 85 per cent of shareholders voted in my favour, the Pongrass brothers were able to come up with over 2,000,000 shares at the poll whereas I only had about 4 per cent of the company, plus those people who voted in my favour, and could only come up with 1.4 million. As a result I had to step down and resign as chairman and director of Stellar Mining and Pongrass took over. So, as is too often the case, with comparatively small companies in Australia, Stellar was now in the hands

Audimco's diamond recovery equipment at Copeton designed by William Joris.

of people with no knowledge of the business concerned. There followed a rapid decline in the company's fortunes. For example when I left the company two granite bars in Reddistone Creek, north of Glen Innes which I realised were the ideal place for sapphires to gather were sold to Wally Graham for $5,000 who I believe made about $3,000,000 out of the deal. All the areas in Western Australia were allowed to lapse. Subsequently C.R.A. finished up with Ellendale, which as I mentioned earlier, has produced some of the most beautiful stones I have ever seen.

As far as I know, the only place that was kept on was Horse Gully, which produced sapphires until quite recently but in fact production was stopped for some reason. One day I mentioned to David Lane, who was a director of Stellar, "You're mad letting it go, because you have pushed a lot of wash through the trommels much too fast. There is a limit to what the machinery can take and you finish up by pushing it in one end and out the other. Whereas if you go through that area again, I think you will find it is worth around about $5 per yard," and that's exactly what it turned out to be.

It still amazes me that although I mentioned to the board of directors at two different meetings of Stellar that I had been invited to become chairman of Audimco, nobody seemed to object, and nobody seemed to worry until a year later when "Conflict of Interest" was bandied about.

Here is an article which appeared in the Daily Mirror on Thursday 6th May, 1971:

STELLAR 'CONFLICT' HAS UNIQUE INTEREST BY Finance Editor Geoffrey Short.

On Monday week, Mr. Albert Joris will know whether his services as chairman and a director of Stellar Mining NL are still required.

At the same time, Mr Alexander Pongrass, vice-chairman of Stellar, will know whether his interests and two board nominees have gained control of the company.

And this will be decided by some 7,200 shareholders voting on a motion unique to the Australian business and mining scene but one which could ultimately have widespread ramifications.

DEBATED

The "Conflict of Interests" issue is a line being pursued by the Senate Select Committee on Securities and Exchange, which should make some interesting recommendations in its report.

Meantime, the Stellar affair is being keenly followed and debated — for what there is to go on.

Simply, Mr Pongrass has charged Mr. Joris with a "conflict of interests" and wants him removed from the board. Mr. Joris says he has no conflict of interest.

To be fair, both Mr. Joris and Mr. Pongrass have endeavoured to keep the basis of this unprecedented dispute as straightforward as possible.

But it is still very confusing and even ironic in that what they are clashing over may never be found — diamonds.

Up near Inverell in the New England district, a new exploration company called Audimco Ltd., has taken out an exploration licence to search for Group One minerals, (tin, zircon, etc) in an area overlapping by 370 square miles an area held by Stellar in which it is searching for diamonds (Group seven minerals).

It so happens Mr. Joris is chairman of Audimco, and a major shareholder and this Mr. Pongrass claims is "highly undesirable" because Audimco's main intention, as stated in its prospectus, is to search for diamonds in the New England district.

Mr. Joris, a diamond merchant, replied that a year ago he made an immediate and full disclosure of his Audimco interest to the Stellar board, he had refrained from voting on or discussing matters involving Audimco, and that two Queen's Counsels in separate opinions thought that there was no conflict, "but that my continued association with both companies could be of distinct benefit to Stellar".

That in a nutshell is where the matter has been let rest.

But several other points should be clarified as best as possible before the meeting.

If Mr. Joris notified the Stellar board of his Audimco interest more that a year ago, why has the matter only now been brought up?

ELECTED

Why was Mr. Joris elected by directors, including Mr. Pongrass, as Stellar chairman to succeed Sir William Walkley on his retirement from the board earlier this year?

If Audimco finds diamonds in its search for minerals covered by its Group One exploration licence, can it then apply for a mineral lease to mine the diamonds?

(Opinion in the city conflicts over this point. Many experienced mining men said there was no precedent to go by.)

As diamonds are a very specialised field, why is Mr. Pongrass endeavouring to replace Mr. Joris with a businessman, Mr. George Pongrass, and also have elected a uranium geologist, Mr. J. R. Stewart, who is executive director of Kratos Uranium Ltd., of which the Pongrass brothers are directors and large shareholders?

Do Mr. Pongrass, who controls some 30 per cent of Stellar, and Mr. Joris, about 4 per cent, intend to refrain from voting until the feelings of the (7200) remaining shareholders are obvious?

Mr. Pongrass only requires a 51 per cent majority to succeed.

In these days of farm-ins, farm-outs, and what have you, was some rational attempt made to settle the somewhat tenuous differences between the two companies on this matter, particularly as both have other extensive exploration areas?

After all, as one mining man put it, the search for diamonds is like looking for a needle in a haystack — only you have to find the haystack first.

The questions could go on.

But the opinion of Sir William Walkley, who sponsored both Mr. Pongrass and Mr. Joris to join the board is interesting, if not enlightening.

"I didn't agree with either of them," Sir William said, "so I sold up and got out".

....

Another article followed:

DAILY MIRROR, MIRROMONEY, 18th MAY, 1971.
EVEN THE STELLAR VOTING CONFLICTED

Talking Trade with Brian Mahoney

The full details of voting at Stellar Mining's extraordinary general meeting which removed Albert Joris from the board are very interesting.

They indicate typical shareholder apathy, and a more interesting point — that comparatively few shareholders were convinced Joris had a possible "conflict of interests." (Although on the surface his defeat by 2.1 million votes to 1.4 million appeared to prove the opposite.)

First, the apathy — only 34 per cent of the shareholders with the right to vote actually voted . . . 1530 out of a possible 5250 or so.

But 400 of the 1530 might as well not have bothered — their votes were declared invalid for one reason or another. (A hundred didn't get their proxies in on time.)

Of the 1130 shareholders whose votes were counted, 917 were in favour of keeping Mr. Joris on the board, a big contrast to the 213 voting against him.

This was indicated in the show-of-hands vote taken at the meeting — 40 supporting Joris and 14 against. But the poll, including the proxy votes, dramatically reversed the decision — 2.1 million votes against 1.4 million.

If it had been based on the number of shareholders instead of number of votes Joris would still be on the board.

But as it was although Alex Pongrass had less "friends" he owned or had interests which controlled bigger parcels of shares.

The whole exercise seems to indicate that not many Stellar shareholders were convinced of a "conflict of interests" by Albert Joris being on the boards of two diamond explorers at the one time Stellar and Audimco.

And about the apathy — Stellar's accountants say the participation at the extraordinary general meeting was "good" compared with the usual number of votes cast at an ordinary meeting.

Which makes it appear that shareholders in mineral exploration companies at least are more concerned with what the share price is doing rather than what the directors are doing!

CHAPTER ELEVEN

An Australian Dream

A friend of mine, Fred Messara, had befriended a French geologist by the name of Jean Casteran, who had had a number of years diamond exploration experience in West Africa. Casteran had also been manager of a subsidiary company of the French Bureau of Geology and Minerals, which is a semi-Government organisation, and had travelled throughout Western Australia in search of diamonds. Fred Messara asked me to join on the board of Audimco as chairman, with himself as deputy chairman, and together with Mr. Casteran, Mr. John Denis an engineer with Citro, and Mr. Graham McColl as secretary, who was later to become a director of the company. The company was floated at such a difficult time, that we couldn't even get an underwriter. We had a broker, Ralph King & Yuill. I think a lot of people were rather surprised and wondered how the company had come on stream so well.

About 14 mining tenaments were brought from Table Minerals Development Pty. Limited, a further 5 exploration licences were applied for in the Bingara/Copeton area south of Inverell and an option was also held over four other areas. Later 2 authorities to prospect over 3,500 square miles were applied for and granted in the south Tanami area in the Northern Territory. I still regret the fact that nothing was done about these areas in the Northern Territory. In any case equipment was bought and shipped up to Bingara and work commenced. A few diamonds were found there, though not very large, they were good quality coloured stones. Bingara, like Copeton, had produced diamonds at the turn of the century, and the old Monte Christo mine was one of the areas we looked at.

Some time later I suggested to Casteran that he come on a trip with me to Copeton. I showed him where, years ago, a small syndicate made up of my late father, Reg O'Brien, Eric Miller, Q.C. and myself had been partners in working the Old Farm Mine. I explained to him how we had sunk a shaft and then a tunnel, which kept dipping down towards a granite bar and how water had stopped us at that particular time. I suggested to him that if he sunk a shaft on the other side of the granite bar on top of the hill and went down to the bottom, that he might very well find diamonds and tin. Before doing this, of course, I advised that we should do some diamond drilling. Seven holes were put down in the area, which confirmed that fact that there was indeed not one, but two layers of basalt, which had formed over that area at different times.

An inclined adit was put down from the side of the hill, down towards the centre of it and whilst it encountered some interesting lignite very few diamonds were recovered. The idea was to get to the lowest point of the depressions, anything below 560 metres being of interest.

As I knew Sir James Foot of Mount Isa Mines, I contacted him and they became interested in doing a joint venture with Audimco. I subsequently instructed Tim Bennett, who was the head of their exploration arm called Carpenteria Exploration, to send men down to work there with Casteran. We had David Crab, who was then stationed at Bathurst to look after the interest at Copeton. Bill Murray was in charge of Eastern Australia, ably supported by George Lyle who did quite a bit of work there, particularly on the recovery plant. After completion of the planned drilling programme, Mount Isa Mines decided to withdraw from the joint venture and Audimco acquired the interests of M.I.M.

My son Bill Joris, who is a mining engineer also, had quite a lot to do with the design of the plant. Bill had some practical experience in diamond mining during his holidays when he was at university with the De Beers Company at Kimberley in South Africa. In July, 1976 the company entered a joint venture with Sibeka of Belgium and Hambros Bank, with each contributing $100,000. This money was to be spent on underground testing, below 565 metres. A shaft had been sunk to the bottom and extensive tunnelling had taken place from there. Then at various spaces where the bottom dipped, winzes were put down. However, both Sibeka and Hambros withdrew from the joint venture prior to completion of the winzes. Although I had known Copeton and for that matter the Inverell districts, ever since we came to Australia in 1930, or just after that anyway, I had never been terribly enthusiastic about finding the source of the Copeton or the Inverell diamonds. There is no doubt, however, that if this source is ever found, it would prove to be a fabulously rich find. But with all the upheaval that has taken place geologically and with the layers of basalt covering the area, the source is going to be extremely difficult to find.

I was always much more interested in going to Western Australia. I tried to talk Jean Casteran and the board into going to Western Australia, but neither Casteran nor the board were very anxious to move away from Copeton, the excuse being, of course, that it was too expensive. I know that it is very expensive to work, especially in remote areas and I also realise that you can't be in ten different places at the same time. But it seemed to me that they had made up their minds to battle on at Copeton no matter what. I decided to resign as chairman of the company.

As far as Copeton diamonds are concerned if ever diamonds are found there in large quantities, I strongly suggest that they not be sold through the Central Selling Organisation, my reason for saying this is that the Copeton diamonds are completely different to any other industrial diamonds. They have special qualities and should be sold on their own merits. The Central Selling Organisation would merely toss them in with diamonds from other parts of the world and the Australian Copeton diamond would completely lose its identity. Only about 5 per cent of these diamonds can be cut into gems. The others are too hard and with wages what they are today it would be impossible and impracticable to do it, except for certain industrial purposes, for which they are especially suited. Apart from being extremely hard, they don't graphitize like other diamonds do, at from 750 to 850 degrees centigrade. You can raise these stones up to 1300 to 1400 degrees centigrade and they will hardly show any graphitization at all. They are very heavy on Carbon 13 and it has been said that they don't need an ordinary kimberlitic environment for them to be formed. I very much doubt if this is true.

As these diamonds have been found on both sides of the Dividing Range, not only at Copeton south of Inverell, but also at Port Macquarie on the coast, it leads me to believe, that the real source of these diamonds could be much further eastward than that area being looked at at present. To hazard a guess, I would be more inclined to have a good look around Guyra, which is south of Glen Innes. Tony MacNevin, who is a very prominent diamond geologist, formerly with the Mines Department, agrees with me on this point.

Apart from the Copeton diamonds, there are as far as I know two other type of diamonds in New South Wales. They are not hard like the Inverell diamonds, and most of them are suitable for cutting into gems. The source of these diamonds is waiting to be found, but I wouldn't be at all surprised if it were eventually found in the region — north of Capertee around Mudgee and the Cudgegong River to Gulgong and across to Wellington and possibly even further north to Coolah. So there is still a lot of searching to be done in New South Wales.

In Western Australia, apart from the 25 carat diamond found on the body of George Stansmore, a member of the 1896 David Carnegie expedition across the Simpson Desert, my father had heard of diamonds in the far north of Western Australia.

After Angus & Robertson published "Stone of Destiny" written by Jack Idriess in 1948, based on the notes he had been sent by my late father, we regularly used to receive matchboxes with various stones in them, from people in Western Australia, who thought they might be diamonds. Invariably they were not. We pounced upon the boxes whenever they arrived and opened them, only to be disappointed by the fact that they were not diamonds. My late father also used to correspond with a Belgian monk, who was at the Kalumburu Forest River Mission, a Spanish Mission Station at the very northern top of Western Australia, the monk told my father that there were diamonds in the area, but also added "Forget about it, because you will never be able to do anything about it, it is all sacred Aboriginal grounds." However, it was in that very area, in 1973, that a consortium, consisting of Northern Mining, Sibeka, London Tin (later to become a Malaysian Mining Corporation), Jennings Industries and Tanganika Holdings, first found a minute diamond. And in late November, 1975 Rees Towie of Northern Mining, sought a joint venture with C.R.A. (Ashton Joint Venture), which resulted in C.R.A. becoming the managers of the exploration. In 1977 a large group of kimberlite pipes was defined at Ellendale, which is 120 kilometres east of Derby and approximately 2,000 kilometres from Perth. A sampling plant was commissioned in 1978 and further work was carried out during 1979 and eary 1980. Some thirty kimberlite bodies were sampled at Ellendale, but out of that lot only two gave positive indications of containing diamonds in commercial quantities. Pipe A has a surface area of 46 hectares and Pipe B 86 hectares.

I have seen these diamonds, which impressed me greatly. They rank amongst the most beautiful crystals I have ever seen, with some really fancy colours amongst them. In many ways they are similar to the diamonds found by John Williamson in Tanzania.

Late in 1979, the Argyle diamond deposit was discovered south of Lake Argyle which is near the Northern Territory border about 600 kilometres from Ellendale and 2,200 kilometres from Perth. Another was the discovery of diamonds at Smoke Creek, with the discovery of kimberlite pipes there. Diamond pipe AKI has a surface area of 45 hectares, and alluvial diamonds have been recovered along Smoke Creek up to 30 kilometres from the pipe. Heavy media separation plants, similar to those at Ellendale are being used at Argyle, and it is rather interesting to hear that the heavy media separation plant which I ordered for Stellar Mining and which was bought under great protests from some of the other directors, is still being used in Western Australia and is being hired out to diamond prospectors. The plant did an excellent job in recovering sapphires at Horse Gully and now apparently is doing an excellent job doing the same sort of thing with diamonds in Western Australia. It was probably one of the first plants built by Mitchell Cotts in Australia.

Upper Smoke Creek is in the valley immediately adjacent to the kimberlite pipe and apparently contains an average grade of 8 carats per cubic metre, which is enormous, and it has been assessed by the Ashton Joint Venture as containing 15 per cent gem quality diamonds. A great problem for the Ashton Joint Venture was having to deal with a number of Aboriginal communities living in that particular area. These communities number 35 to about 250 and they had to deal with the Aboriginal land trust as well. It is to the great credit of the field staff of the Ashton Joint Venture that they were able to satisfy all concerned. The main object was of course to minimise, as far as possible, the possible disturbance to Aboriginal sites and communities. It appears that the Ashton Joint Venture field personnel have done a wonderful job in maintaining significant Aboriginal sacred sites. Some of these sites, apparently date from so many hundreds of years ago that some present-day Aborigines didn't even know of their existence. There was also a sweetener on the part of the Ashton Joint Venture for the Mandangala Community who live at Glen Hill Station which comes under the aegis of the Aboriginal Land Trust. In this case the Ashton Joint Venture agreed to improve and upgrade roads, erect fences, provide buildings and services, station vehicles, trucks, supply electric power from a station type generator, drill a water bore, and supply a windmill and pump in addition to providing materials and assistance for the erection of stockyards. Approximately $200,000 will be spent over a 23 month period to do all this and after that there will be $100,000 paid every year, this being indexed for inflation. It is pleasing to note that the Ashton Joint Venture, through the management of C.R.A., have done such a wonderful job in creating a friendly atmosphere in dealings with the local Aborigines, who will no doubt benefit from all the help and assistance they are getting.

As for my own involvement with West Australian diamonds, apart from what I have already said earlier, I was once approached by an officer of C.R.A., who had been in Bougainville before that, and who had a great number of our diamond drill bits. He wanted to know what I could tell them about diamonds, and whether I could give them any books, or information on diamonds, which I did. I became a little bit curious and asked him what he was doing. He answered that the board of C.R.A. had given a number of officers various jobs to do. Each had to do a feasibility study with one covering copper, another nickel, another gold, and his job was to do a feasibility study on diamonds. Then in early 1980 I was visited by two C.R.A. officers from Perth, who came over to discuss the possibility of marketing Australian diamonds from Western Australia. I explained, in detail, the four different methods they could use to market them. I also mentioned the letter I had received from the Right Honourable Douglas Anthony, then Deputy Prime Minister, dated 27th November, 1978. I have no doubt that this same letter was sent to a number of people. Principally the letter advocated more than just the cutting and polishing of precious stones in Australia. In other words value adding to the rough material, rather than having this done outside Australia. I forwarded a lengthy and extremely detailed reply and then apparently various committees were set up to investigate the matter. Subsequently I corresponded with Mr. J.T. Starkey of the Department of National Development and Energy, and then later with Mr. R.B. Bolduan of the Department of Trade and Resouces. I also dealt with a Mr. L.W. Owens of the Department of Mines and Energy, South Australia, and Mr P. Ryan, of the Commonwealth/State Joint Study Group on Raw Materials Processing, Department of National Development and Energy. The thing that struck me most about all this was the fact that a committee had been set up consisting mainly of people who knew nothing about the diamond business and who hopefully wrote lengthy reports and corresponded with people like me in order to extract information on which no action was taken.

What I had advocated in these communications was an export duty on diamonds, sapphires and opals for the following reasons. First, Australia produces about 90 per cent of the world's opals, about 65 to 70 per cent of the world's sapphires, and very shortly well in excess of 50 per cent of the world's diamonds. I argued that the payment of an export duty would yield revenue or would encourage the owners of mined diamonds to cut them in Australia, Second, if you want to add value to these three precious stones then, the only way to do it is to cause them to be cut and polished in Australia, by imposing an export duty on rough stones.

When in 1929 my father decided to go the South Africa, to set-up a diamond cutting industry, the South African Government already had an export duty on rough diamonds in order to encourage people to set-up diamond cutting facilities there. One could well ask how can we in Australia compete, for example with the low wages paid in Thailand or in India or in a number of other countries in the Far East. Today, nearly all our sapphires go to Bangkok, where they are cut by people earning about $8 to $10 per week. When they are brought back to Australia no duty is imposed on them (or there was a 2 per cent duty, which nobody paid anyway because the goods came back in people's pockets). To compete against such countries is virtually impossible in Australia where you have got to pay wages in excess of $400 per week. This is normally for a 35 to 38 hour week, whereas people in the countries mentioned previously work from sunrise to sundown for a fraction of that amount.

If one takes an average wage cost for a top-class cutter in India, it would be about $600 per year, in Israel $12,000 per year, in Belgium $25,000 and New York $30,000. However, despite the difference in wages between

Belgium and New York quite a few thousand people are employed in New York cutting the bigger and better stones. In India there are over 150,000 cutters, in Israel about 10,000, in Belgium of the 20,000 or so people engaged in the diamond industry, 9,000 to 10,000 are cutters, and in New York there are about 3,000. In other words what I am saying is that if the bigger and better quality stones were cut in Australia, then the small and less valuable stones could be exported, in the case of diamonds to India, and in the case of sapphires to Bangkok for this to be done.

I also advocated a duty on cut stones. There used to be a duty of 10 per cent on diamonds, which was removed a number of years ago. The other major thing I advocated, was the adoption of the *Precious Stones Act* of South Africa. *The Precious Stones Act* in South Africa has been tested and tried over the past fifty-odd years and has been most successful. Briefly what it does is to set-up a system whereby the government can determine how many carats of diamonds may be produced from time to time in order to prevent over-production.

To deal in diamonds or precious stones for that matter, one must have a licence. To get a licence the person concerned has to be a resident of the country for a least three years and have no criminal convictions. In addition the person concerned must lodge a monetary performance bond and pass a practical and oral test to establish that he has sufficient knowledge of the diamond trade to work successfully in it. Purchasing may only be done between sunrise and sundown. A name plate indicating that one is a licensed diamond dealer must be displayed outside one's office and when buying is conducted a dealer must have a small office there and put up a flag with his name on it when he is buying. Furthermore, a dealer must keep records of any stones he buys, from whom he buys them, the number of stones and their weight and if they are specified stones he has to keep complete details of the colour and the impurities if any. In other words all indicators that could lead to the diamond's identification. This sort of thing, if acted upon, must be done on the Federal level in order to make it apply to all States and Territories so as to allow no loophole and it must apply to the three main precious stones that are being produced in Australia in large quantities namely opals, sapphires and diamonds. In South Africa, anyone caught with rough diamonds, and not licensed can go to gaol for up to three years. This helps to keep out undesirable characters. Surely we don't want to finish up like Sierra Leone, where the Lebanese Mafia control the diamond trade, where illicit diamond dealing is rife and where people's throats are cut for small quantities of diamonds. I have advocated the adoption of these measures time and time again to numerous people, but nothing has been done, except for legislation enacted in Western Australia in the *Diamond Ashton Joint Venture Agreement Act* of 1981. Part IV of the Act relates to the security of the diamond mining and processing areas. The Act provides that the Governor may by order in Council declare any land or premises where specified operations are being conducted for the purposes of the agreement to be designated areas for the purposes of Part IV. In this part special provisions with respect to security which apply within a designated area are set out. It is an offence for a person within a designated area to have — without lawful authority or excuse — an uncut diamond in his possession.

Control of entry to and egress from a designated area is provided for in the Act. A person may enter or exit from a designated area only by way of a controlled access point and the Act provides that no person may enter a designated area without the permission of a Security Officer, (who must be the holder of a licence as a guard under the *Security Agents Act*, 1976).

As I see it, under present legislation you can't possess any rough diamonds and you can be fined $10,000 or go to gaol for a couple of years, if you are in possession of rough diamonds in an area where diamonds are being mined. But there is nothing to stop anybody from having rough diamonds on him, in Perth or in Adelaide, in Melbourne, in Sydney, or anywhere else for that matter. Take the case of sapphires, for instance. Any Thai can come to Australia with a briefcase full of dollar notes which he can buy comparatively cheaply in Hong Kong. He can then go out to Inverell, in New South Wales, or to Rubyvale in Queensland, advertise the fact that he is there to buy sapphires and wait for sapphire sellers to come to him. No record is kept. He pays cash for the goods that he buys and no receipt is issued (in many cases the sapphires are stolen). The vendor or the thief selling the sapphires doesn't pay any tax on the money earned. The Thai then takes them to Bangkok and has them cut for sale world-wide. Quite a lot of them are brought back into Australia, in his pockets, so he can avoid paying 2 per cent duty on them and he then sells them, no sales tax added for cash and so the cycle continues. The same happens with opals.

I remember a case, when I was with Arrawatta, sapphires of a day's production (amounting to about $8,000) being stolen. The thieves put a young boy of very slim build up the shute of the pulsator in the middle of the night. The thieves told him what to do and he scraped the sapphire bearing wash out of it. They were eventually caught. They had sold them in a pub for $800 because the people who bought them obviously knew that they hadn't produced them and that they were stolen. They were paid cash and when the thieves were eventually caught, they were let off on a good behaviour bond. Later the same thieves went to Queensland and were holding people up at gunpoint at Rubyvale when they were caught, and taken to prison.

Again in South Africa if you are a licensed diamond dealer and you export rough diamonds you must take the parcel, unsealed, to a police department, which includes diamond experts. Here the diamonds are weighed, examined and valued. As the police concerned are experts in diamonds there is no room for evasion of duty through lower-value invoicing and furthermore since the origin of stones can be identified there is little room for the exploiting of illicity obtained stones. This sort of system could be set-up very easily, and cheaply, here in Australia. I could train a number of young people who could do this sort of thing on a part-time basis, and who could be paid by the hour. But, like everything else, anything is easy, providing you know how to go about it and providing you know what it is all about.

As mentioned earlier I had discussions with two officers from C.R.A. about the possibilities and methods of marketing Australian diamonds. I asked if I could see the diamonds and it was agreed that I could go to Perth and have a look at them. So, in June, 1980 I went to Perth and I was shown the diamonds from both Ellendale and Lake Argyle. The Ellendale diamonds were absolutely magnificent, and were such beautiful specimens that they made my mouth water. The Lake Argyle diamonds were also very interesting. They were a bit rugged, but those of them that were white were top quality in the D,E, and F colours, the best you can get. Then they mentioned to me that they would like a detailed valuation of the goods by sorting and grading out the stones into gem diamonds and industrial diamonds in their various classifications. Of course, I was quite prepared to do this for them but I heard nothing more about the matter.

Early in January, 1981, I was approached by Rees Towie, who invited me to visit Perth with him and value Western Australia's diamonds. Rees of course knew that I had practical experience in the manufacture and dealing in both gem and industrial diamonds. I had met Rees once before and I had been very impressed with the man. He is from Western Australia and has a university degree in engineering. He has spent many years in Canada oil well drilling and contracting and was the chairman and managing director of Northern Mining Corporation. It was, after all, Rees Towie's son John and his wife Maureen (she works under her maiden name of Muggeridge and is a niece of Malcolm) who first found diamonds at Lake Argyle.

Whilst I was there, I sorted 6287½ carats of which about 664 carats were from Ellendale. The percentages of the stones that I obtained were 17.84 per cent of boart, 38.6 per cent of drilling goods (i.e. various sizes, from small drilling material to oil bit sizes, including emery wheel dressers) 28.9 per cent industrial stones and 14.71 per cent cuttables. In grading diamonds for gem quality, as far as colour is concerned the Gemmological Institute of America grading is used which is based on the alphabet. A,B, and C are not used, because a lot of jewellers used to grade their diamond colours in A,B, and C. Some even took it further and went to A1,2,3,B1,2,3,C1,2,3, etc. So the top colour in a diamond is D,E, and F, all the way down to X,Y,Z. D,E, and F, are the top colours and I noticed a lot of the gem quality stones I saw from Lake Argyle were in that category.

That's why, in June, 1981, I, together with several other people from Antwerp, New York and Japan were asked to do a valuation. I personally sorted and graded 16,500 carats.

In order to arrive at the value of the gem stones I had the latest prices from very large companies in Antwerp and London. I didn't use the London or the New York prices as much as the Antwerp prices which were lower. I took a yield of 40 per cent, (which is very low) and calculated cutting costs as $65 per carat. I then reduced the Antwerp price by 25 per cent to arrive at the cost of the rough. But, even then I reduced the colour grading 1 to 2 grades, which greatly reduces the prices of the finished material.

In relation to price one thing that had fascinated me was the replies given by the Minister of Resources Mr. Peter Jones to questions put to him in the Legislative Assembly on the 14th September 1982. Mr. Jones stated that an investment of an additional $350,000,000 in January, 1983 would be required for 2.9 million tonne per annum of kimberlite to be mined. He went on to state that an area, which is known as AKI would produce in the order of 7.5 carats per tonne, with some parts of that area yielding in excess of 10 carats per tonne. In the Upper Smoke Creek he mentioned an average of 8 carats per tonne. If you take 8 carats as an average and you multiply 2.9 million tonne by 8 carats, you come to 23.2 million carats of diamonds, which is about the figure that I have always considered the production would be. I always guessed that production would be between 23 and 25 million carats.

The Minister went on to state that annual revenue would be about \$130,000,000. If you divide 23.2 million carats into \$130,000,000, you come up with an answer of \$5.60 per carat (i.e. gems, industrial diamonds, crushing boart, the whole lot). *

From Western Australia's point of view, the sad part is that of the estimated \$130,000,000 revenue mentioned by the Minister, the State gets only a 7½% royalty. It means that the West Australian Government will get \$9,750,000 per annum in royalties. A friend of mine in England offered to take up the whole production at \$18 per carat and when a number of Indians were out here representing the Indian Government, with Indian Government backing, they offered \$18 per carat also. So, if you take this figure of \$18 per carat and multiply it by 23.2 million carats, you come up with a gross income of \$417.6 million, of which the West Australian Government would get 7½ per cent of royalty, that is \$31.32 million per year. However, on the basis of the answer to question 1301 of the Legislative Assembly of 14th September, 1982, the minimum estimated royalty due to the State is 7½ per cent of \$130,000,000 or \$9.75 million per annum.

Sorting is easy and can be taught to anybody with a tiny bit of common sense and good eyesight in a matter of days. Putting a value on the stone, well that's a different story. That requires a deep knowledge of diamonds both for gems and for industrial purposes. The other alternative is to have one man who is a gem expert, and the other who is an industrial expert, with one doing the gems, and the other the industrials. Also you have one overseer, who has a knowledge of both gems and industrials. Such people being few and far between.

As for doing valuations, for both gems and industrials, this again can be taught to young people. I would select a number of young people who have good keen eyesight, who have done a Gemmological Course and preferably a Diamond Course in Los Angeles. Quite a number of Australians are doing this by correspondence. Just imagine what a tremendous boost it would be for Australia, if there was an export duty on not only diamonds by also opals and sapphires and an import duty on cut goods coming into the country. Just imagine the work it would create, particularly if industrial diamond tools also were tackled with a 15 per cent advantage. It would mean billions of dollars to Australia. This, as I said earlier, can easily be done.

I for one am just one of those people who, although not born in Australia, after 56 years has come to love this country, and its people. It may well be that I'm more Australian than most born Australians and so I hate to see this country being used merely as a quarry for the benefit of others.

One of those things that has puzzled me all along is the fact that both the Ashton Joint Venture and the West Australian and Federal Governments have employed two supposedly independent valuers. It has always been my

* This price in my opinion is extremely low.

opinion that any individual or firm which has an association or is to some extent dependent upon the Central Selling Organisation cannot be a truly independent valuer. The same applies if diamonds not bought by the Central Selling Organisation are sold to a company which to some extent is dependent upon the Central Selling Organisation for its raw material.

So, I've said my piece for what it's worth, in all sincerity. As I have said before, everything is easy, if you know what you are doing.